Ernest Shackleton

The 'Quest' with Ernest Shackleton in command, passing under the tower bridge of 'The Thames' London

Published by
Reardon Publishing
PO Box 919, Cheltenham, UK.

www.antarcticbookshop.com
2nd Edition 2023

Written by
George Henry Ross

ISBN 9781901037210

The 'Quest' on the way to the sea, passing the tower of London

Shackleton's cabin on the Quest by Bill Payne

Ernest Shackleton

Foreword

I have designed and published many books over the years covering the adventures and daring-do of early Antarctic explorers, and it is always a pleasure to find something new to say on the subject.

The death of Ernest Shackleton has of course been covered in many books and one of the best being by Commander Frank Wild, of course we do not expect to compete against his book “Shackleton’s Last Voyage” but instead show the expedition from a different point of view and that being from a lowly below decks seaman by the name of George Ross.

George joined the Quest Expedition, at a place called Leith Harbour, in South Georgia, with the position he writes as donkeyman (a person in charge of a ships engine room) and he explains the daily running of the ship after Shackleton’s death as a first hand account from a below decks point of view.

George Ross’s account of The Shackleton-Rowett Expedition on the Quest is almost completely unknown and I have not changed or amended any of his recollections or words so you will be able to read just what he wrote, and feel how it was to be part a golden age of Antarctic exploration.

Thanks must go to all those which have helped with the background creation of this book and which this book would not exist without their input and help.

First must be my thanks to **Mark McClean** who pointed me in the direction of the mostly unknown George Ross and helped me obtain the manuscript.

Jan Chojecki for our long chats that got me interested in the Quest Expedition. **Jon Sandison** for permission to reproduce his poll bearers' article, along with a short history of Shackleton's other expeditions, **Alan Noake** on the Shackleton Scouts and **Glenn M. Stein,** for images and history of Dr. L.D.A. Hussey's full-size medal group, **Bill Payne** for his Quest linocuts.

Not forgetting The photo libraries based in Australia and New Zealand which along with **Val Kerr, Kay May, Paul Clammer, Ann Lane, Rosie Needham, Geraldine Wythe, Pippa Simpson, Shaw, Dave Connor, Paul Firth, Stephen Scott-Fawcett, Branko Radovinovic, Peter McCarthy,** and **David Wilson** for his insight on Shackleton's Nimrod Expedition whilst working with him on his book Nimrod Illustrated.

Plus many others have helped bring this book alive, with information and images of the people and places covered in this history.

Nimrod Illustrated, Hardback Edition ISBN: 9781873877906
Special Limited Edition ISBN: 9781873877913

Cover photos: Grytviken Church and Shackletons Grave.

N.Reardon

Linocut on Shackleton's Quest
by Bill Payne

INDEX

By courtesy of Illustrated London News

A DIAGRAMMATIC VIEW OF THE *QUEST*

1, Crew's Nest with Gyro-compass; 2, Mark Buoy; 3, Sperry Gyro-compass; 4, Hydrographic Room: 5, Sir Ernest Shackleton's Quarters; 6, Clear View Screen; 7, Kipling's "If"; 8, Semaphore; 9, Range Finder; 10, Standard Binnacle; 11, Meteorological Screen; 12, Gyro-compass; 13, Wireless Room; 14, Life-boat Deck; 15, One of two Life-boats; 16, Mark Buoy; 17, Water Tank; 18, Kelvin Sounding Machine; 19, Surf Boat; 20, Stowage for Stores and Specimens; 21, Sleeping Accommodation for Naturalist and Photographer; 22, Windlass; 23, Dark-room; 24, Chain-locker; 25, Lucas Sounding Machine; 26, Stores; 27, 15-ton Water Tank; 28 and 29, Stores; 30, High-power Wireless Room; 31, Coal Bunkers; 32, Boiler; 33, Galley; 34, Avro; 35, Main Engines; 36, Engine Room; 37, Ward Room,

CHAPTER 1.

READERS, I am writing this story to try to interest you in my experiences whilst I was serving as donkey-man on board the "*QUEST,* R.Y.S.," with the Shackleton-Rowett Expedition.

Quest

The "*Quest*" was commonly known as the Little Wonder Ship, and that she turned out to be, in the end. I joined her on January lst, 1921, at a place called Leith Harbour, in South Georgia, Which is one of the islands where the Whaling Industry is carried on. It was also one of the last we saw before going into the Unknown.

The "*Quest*" had called at this island for bunkers, and also to make the final preparations before we made our dash for our destination in the Antarctic Regions.

South Georgia

It was at Gretviken, South Georgia, that our gallant leader, Sir Ernest Shackleton, passed away. It was at first arranged for his body to be sent home to England, but these orders were cancelled, as it was found that our leader's wish had 'been to be buried in the Antarctic Regions. So, in fulfilment of his wish, he was interred at Gretviken, his remains being committed to the earth while we were ploughing our way into the ice.

The bearers who had the great honour of carrying our gallant leader to his last resting place, were six Shetland men, employed by Salveston & Co. of Leith, Scotland, in their Antarctic Whaling Industry; while the Governor of the island, together with one of our doctors, who was left behind to superintend the burial arrangements, officiated at the ceremony. After the death of Sir Ernest Shackleton, Commander Wild took over the command of the expedition. It was he who gave me the berth aboard the *"Quest."* After telling me of the different hardships one meets with on a perilous trip like that which the *"Quest"* was about to make, he asked me if 1 still wanted to go and of course I replied that I did; and so after the agreement had been read out to me, I was duly signed on as a member of the Expedition. We signed on for three years, or longer if required.

The Whaling Station at Grytviken

I joined the boat the same day, and was very excited. I was at once given my Antarctic clothing, and received plenty to keep me; and me boys warm, who were a fine lot of fellows, the majority of whom had been on smaller expeditions before, soon made me feel at home.

One of the crew had been on three other such Expeditions, and when we were at leisure he would tell us of some of his very exciting experiences; and he also gave us plenty of advice which came in very handy before our voyage was at an end.

The Wireless Operating Room

When I had been shown my duties in the engine room , I soon begun to feel at ease, as there was plenty to do.

The *"Quest"* R.Y.S., a ship of 180 Tons Register, was both sail and stream. She sailed under the Royal Yacht Squadron, and was fitted with all the latest equipment.

We had two sets of wireless on board, one being for emergency; and we had two dynamos, one for the wireless, and the other for the lighting of the ship, and for the different electrically driven machines such as the sounding machines, and others which I will explain later.

Kerr (Chief Engineer) Examining
The Lucas Deep Sea Sounding Machine

Alexander John Henry Kerr

Val Kerr

George Ross looking towards camera during a deep sea sounding

There were two sounding machines and both were of the Kelvin make. One took up to 500 fathoms and the other, the deep water one, up to 5,000 fathoms.

The biggest sounding I saw taken was one of 4,000 fathoms. This machine was easily worked. The lead was put over the side, with weights attached to the wire in such a way that, when they touched bottom they were released, and thus indicated the depth on the clock which was on the machine. As the wire was running out a man stood by with a hook on the wire, and holding the brake of the machine and then as he felt jar of the weights as they touched the bottom, he put on the brake. The weights used on the wire were always lost; but we had a great number of them on board.

We also had on board a very small flying–machine a "Baby Avro." But this was used practically only for taking photos and for observations of the ice. This machine was easily stowed, the motor-engine being lashed under the boat-deck and the wings unshipped and put away.

"Baby Avro"

"Baby Avro"

The Enclosed Bridge of the Quest

We had clear "Clear Screen Ports" fitted up on the bridge, these being used mostly when there was a Snow blizzard. They were equipped with electric fittings, and were driven around at a terrific speed, this keeping a clear view for navigation. These ports helped us through many a big blizzard, and were a great help.

There were two life-boats. They were splendid boats, and could, if required, be turned into sledges. The forepart of the ship was fitted out for the geologist and naturalist, who did all their work there.

The Schermuly Portable Apparatus

Setting up Kites for the taking of Meteorological Observation

The Tow Net in use

The naturalist was also the cinema-operator, and he was kept busy with plenty of work to do.

We had plenty of wholesome food, but the greater part of it was tinned food, such as beef, vegetables, cakes, fish and fruit. At first we got fed up with this kind of food but after a while we got used to it. However we had a good supply of whale beef aboard, and also seal flesh and penguin, which made very tasty meals.

Now that I have explained all about the ship and Her equipment, I will try to interest my readers with an account of our trip to the Antarctic

Quest

Val Kerr

Frank Wild

CHAPTER II.

Heading for the South

It was about the 5th January when we left South Georgia and set a course for the Sandwich Isles, distant about a two days' run. We encountered very heavy seas, and with it blowing very hard, we frequently “shipped it green."

Eventually we arrived off the islands, and they were indeed a splendid sight. The one nearest to us seemed to be literally covered with penguins, which, with the sun shining on their white breasts, made a fine picture.

Our film operator wanted to get a picture of the birds in flight, and gave an order for a shot to be fired in the air as soon as he was ready. This was done, and, then, what a sight ! The birds could not get into the water soon enough.

After taking more films, we proceeded on our way to the south. We now passed plenty of heavy ice, and also sighted some icebergs, and pretty big ones, too. The soundings, taken every hour, ranged from 500 to 1,500 fathoms. We continued to get bad weather and the ship rolled very heavily.

The Quest Escapes an Iceburg

After steaming three days, we encountered very heavy ice, but we still made good headway. We now began to feel the cold intensely, for it was a lot below freezing point - and one feels such a temperature after working in the engine-room!

One morning I was called as usual at 4 a.m. It was always my custom to have a look round, and that morning I got a big surprise. The ship was in one vast field of ice, and one could hardly see any signs of water apart from some huge cracks in the ice. There were also a large number of icebergs in all directions in the distance.

I must tell my readers that this was my first experience of the ice in the Antarctic, and I felt anything but brave and comfortable.

Entering the Pack Ice

Collecting Ice for replenishing the Water Tanks

I had been on plenty of trawlers to Iceland, but that was nothing compared to this kind of life. As time passed on, however, I got used to it; and, indeed, it was nothing compared to what I had yet to experience.

In the afternoon we were told we could go on the ice for exercise, and this we did, and were able to have a game of football by filling a football case with pieces of old cloth. We did this, because, with the wind that was blowing, a ball with a bladder would have been too light, and would have been blown away.

While we were plying the game, one large penguin stood watching us, and every now and then it let out a squeak. It remained there all the time we were playing, except when our mascot saw it and went towards it, when it went away, returning again, however, as soon as the dog had departed.

These penguins put one in mind of some of our noted comedians of today; their walk and behaviour are so comical.

The ice now seemed to be getting still thicker; and one iceberg we passed, was quite close to us, being not more than 100 yards away.

The wake of ice as seen from the Crows Nest.

A close up view of the pack

It made a tremendous roaring and rumbling, just like thunder, this being due to the fact that there was a great hole in the side of it, into which the water kept forcing its way.

And now the ice had so increased, that we only made ten miles in twenty-four hours; and finally we could not make any headway at all, but just drifted where the ice took us. We had plenty of exercise however, to keep us in trim, playing football and various other sports.

It was terribly cold, but, one consolation, we had plenty of warm clothing. We now kept up just enough steam in the boilers to warm the engines through and provide warmth for the living quarters.

In addition, we gave the engines a run every now and then, to keep them from freezing up, as it would not have taken long for them to do that, if we had not run them.

We had plenty of food and amusements; and with a gramophone and about 500 records, which had been presented to the ship before she left London; we were able to pass many a dreary hour away, and to forget our worries for a time, anyway.

Killers rising to blow

The Quest pushing through thin ice

And one cannot help worrying at this stage of such a voyage, for it is the worst part of it.

When our work was finished, we always went on the ice for a game of football, or to look for seals and sea-leopards, trying to get a good specimen. At last we can ht a fine specimen of a sea-leopard, being one of the biggest that our naturalist had seen - and he had seen a good many of them in his time. After it had been killed it was taken on board to be photographed, and then the naturalist prepared it for keeping.

Forging ahead through loose Pack Ice

In the Antarctic: The Quest a mass of Frozen Spray

CHAPTER III.
Blocked in the Ice.

It will now be convenient to me to leave the usual narrative form in telling my tale, and to proceed, for a time, at all events, in the present tense, in which of course I made my notes at the time.

We are blocked in the ice, and it looks pretty bad, as the ice is pressing up the sides of our vessel, and we are anxiously waiting to see how matters will turn out.

The wind is still blowing strong, and plenty of snow is falling. One cannot see any colour but white about the ship, and she herself is covered with snow.

At night it is very difficult to get any sleep, as the noises the ice makes keep one awake. These are due to the ice pressing on, and also rubbing, the sides of the ship. The sound is mostly that of a moan, but occasionally there is a creak which is the pressure of the ice.

Those of us, whose work is in the engine-room, hear these noises more plainly. There are two thick beams in the after-part of the engine-room, which is aft, and every now and then, owing to

The "Quest" Beset

the pressure on these beams, there comes a creak that gives us all a start. However, we are already, as the time goes on, getting used to it.

The boys are giving up hopes of getting out of here this season, but we all have given up worrying about it. We are hoping, however, for plenty of North Gales, for they will help to break up the ice so that we can get the ship turned. Yes we are all wishing for the wind, and if ever a man prayed for wind, I have done so lately. I am now beginning to see the bad parts of this expedition.

We have been doing plenty of cleaning and repairing, and other small jobs that have got to be done.

Finding the Magnetic Dip: Jeffrey and Douglas at work

We have been blocked in the ice for 18 days now, and it is looking just as bad as ever. But we are still hoping to get a good gale of wind, for then we shall stand a chance of getting the ship turned round, and then getting into open water again.

Here is a good wind blowing. In fact, it has latterly turned into a hurricane, and I can assure my readers that it is blowing strong !

One can now lie flat on the ice, and, looking along the surface see the steady heave of it-just a steady swell. Yes, the ice is breaking up in all directions, and in her immediate vicinity, the ship is already clear of it, instead of having it pressing against her sides. But there is no use in our trying to turn the ship, as the gale is still raging.

When we got on deck this the next-morning, all hands got busy to try and get the ship turned, in the hope of then getting her clear of the ice. But after trying two or three times, we had to give it up as a bad job, the wind being too strong for us. We are now going to wait till we get a lull in the weather.

The wind has died down to-night, and we are going to have another go at it ; and I think we shall manage it this time, as the ice has broken up considerably. After a lot of work and trouble, we managed at last to get the ship turned, and now, with our watches set, we are ploughing our way to the open water.

After three days punching, we have eventually got to the open water; but now the wind is blowing so very hard that we are going to remain in the pack till the weather calms down again.

We have received orders from the Commander, that we are going to have another try for the South - we are going to the Weidle Sea.

We are steaming for the Weidle Sea - and we are running out of coal. But as soon as we get a fair wind, we shall set sail, and economise on our coal, of which we have only about 50 tons left.

We are rolling very heavily and dangerously. In fact we seem to be top-heavy, now that we have used most of our coal, which was good ballast.

We have get to keep a good look-out now, as there is a lot of floating ice around us, and it is very dangerous navigating through it. It is still terribly cold, and there is still a good breeze blowing.

We have started to head for the South again, and I hope we shall have better luck this time. I shall be glad when we get in the ice again as then the ship, which is rolling very dangerously, will be a lot steadier.

Pushing South through heavy pack

We are shipping plenty of water down the engine room and stoke-hole, and us poor firemen are stoking in sea-boots, on account of the water running over the foot-plates. We have named the stoke-hole foot-plates, *Brighton Beach,* because there is always plenty of water, and also ashes, there, on account of the ship always rolling.

We are having plenty of trouble with the bilge-pumps, as the small coal and ashes are continually blocking up the pump.

After two days sailing, we have got into the pack-ice again, and begun to punch our way through South once more; but we are making very poor headway, as the wind is against us.

After ploughing our way into the ice for three days, we have had to give up the attempt to get any further, as the ice is getting too thick for us, and the wind is still a head wind for us.

So our Commander is thinking about turning the ship about, as it is absolutely impossible to go any further South.

Our coal supply is down to about twenty tons, and we shall need that before we get out of the ice. We have got the ship about, after a lot of trouble and hard work, and are now steaming for the open water.

After a couple of days we sighted the open water, and, as usual, we had a good breeze to greet us, it later turning into a gale.

We have had to put all the heavy gear which we had on deck, down below, as the ship is rolling very dangerously, and we have only 5 tons of coal left, so that we shall have to save as much as we can to take us into any harbour we can make.

The bilges and pumps are giving us a lot of trouble, and so keeping the engine-room staff busy in keeping the water down.

The Quest pushing North through rapidly freezing ice

CHAPTER IV
Heading For Elephant Isle

We have now headed for Elephant Isle. This island got its name from the fact of there being so many sea-elephants on it ; and when we arrived there, the beach was absolutely crammed with them, the males, as usual, keeping near their mates, to give them warning of danger, and, if possible, to protect them.

These monster seals, which are very ugly-looking creatures, sometimes weigh as much as a ton; but, in spite of their size, they can get over the ground pretty quickly, propelling themselves along with their huge flippers.

After finding a suitable anchorage, we put our Survey Party on shore, and then came back to the ship, as were going to take soundings round the island. There is still a good breeze blowing.

Our cinema-man is keeping busy in taking plenty of good pictures of this place.

The soundings we have taken range from 100 to 400 fathoms. There are a tremendous lot of rocks which make it very bad for navigation.

We are doing all the small repairs needed in the engine-room, while we have the chance.

It was at this island that Sir Ernest Shackleton left stores, and also a life-boat, when he made his last expedition on the *"ENDURANCE"* which was smashed up in the attempt to get to the South; we were unable to reach the place where the boat and stores were left, on account of the heavy seas running.

The wind is freshening, and we have given the signal for the survey party to return to the ship, before it is too late to get them off.

After a lot of trouble we got the survey party off, and then steamed away for our next destination, which will be South Georgia, where we are going to pay our last respects to our late leader, Sir Ernest Shackleton, who gave his life for scientific research work.

After 4 days, we have arrived at Leith Harbour where we are going to bunker, and also do a few repairs both on deck and in the engine-room we shall then proceed to Sir Ernest Shackletons last resting place at Grytviken.

On the way to the cairn - seen in the distance

Looking shorewards from the cairn

Farewell to the Boss

Shackleton's grave at Grytviken

The Cairn erected to the memory of Shackleton

Shackleton's grave on the slope behind the buildings

The Quest in Grytviken Harbour. Moranen Fiord and The Allardyce Range behind

At 7 a.m. this morning, we proceeded on shore at Gretviken, to pay our last respects as mentioned, and also to erect a memorial in the most suitable place. This it was decided, was on the side of a hill, as there the memorial could be seen by any passing ship.

The memorial was a great oak cross, with a suitable inscription on it, supported by large rocks built into a huge cairn. A bottle containing a photograph of all the members of the expedition, and a list of their names, was buried beneath the cross. The memorial stood about 15 feet high.

Grytviken, Cumberland bay, where Sir Ernest Shackleton died.

After taking a film and a number of photos, we returned to the ship.

We are getting ready for our run to the next group of islands on our way to Cape Town.

After three days' bad weather, we arrived at our first place of call, namely Gouth Isle, and after steaming nearly all round the island, we came to a good anchorage, and also a 1anding-place.

Our Commander then gave permission 'for the crew to go ashore but only half at a tune-this on account of the sudden squalls of wind that keep rising.

This island has had the name of diamonds having been found on it, and all hands were excited at the prospect of having a look round it.

When the party I was with, got ashore, the first thing we saw on the beach were the remains of three huts. Inside them were a lot of tins that had contained food, but the weather had rotted them, and they were just falling to pieces.

Outside the huts there was a big hand-pump, and what had been a tray for washing the dirt, to find the diamonds; but it looked as if the attempt had proved a failure, for there was every appearance of the survey party having left in a hurry. They had however; found some diamonds, but these were of such an inferior kind, that they had been discarded.

There were plenty of tunnels where the diggers had been working, but these also bore evidence to the fact that the results had not been worth while.

There were plenty of fine birds here, and the fish was very plentiful, especially lobsters and crawfish. Leaving Gouth Isle, we set our course for an Island called Inaccessible Island, and after three hours' steaming we arrived there.

It looked as if the island got its name from what it was, for when our Survey party went ashore, they found that they had their work cut out to climb up the cliffs, in order to reach the land. Our geologist says it was the hardest climb they have had for a long time.

The island looked very bare and dreary when we got alongside, but our Survey party reported there was good fertile land, with fine grass pasture and plenty of flowers ; and also that there was a fine flat surface.

And though they soon returned, they brought back plenty of specimens.

We have started to catch some of the fish, with which the water here is literally swarming. As soon as we throw anything in the water, there are thousands of fish after it.

These fish are called *Five Finger Fish,* on account of their having marks like the four finger: and thumb on their sides; and in view of the great multitude of them, our scientist has named this fishing ground the Forest of Fish.

There were also plenty of lobsters and of various other kinds of fish; and we secured a fine supply of all that we required before we left the island.

We let two young geese go here, for an experiment, to see if they would breed.

Our next call will be at an island called Tristan da Cunha, on our way to Cape Town.

CHAPTER V.
The Loneliest Island in the World
Tristan Da Cunha.

We are bound for Cape Town, but we are going to call at Tristan da Cunha on our way. We are now in the Roaring Forties, so named as every sailor knows on account of the bad Weather and bad swells that are encountered here at this time of the year.

We are certainly getting very bad weather, and it is raining very heavily, but it is getting very warm, Tristan da Cunha is a very small island, where years ago some shipwrecked seamen and one or two settlers stayed, and tried to make a living. But as the years have gone by, the population has grown to over 150 all told.

We arrived at the island and found a good anchorage, but before we could get our boats lowered, men from ashore came alongside, and said they would put us ashore when we were ready.

The islander's boats were strongly built. Made of canvas, and their ribs consisting of stout branches of trees, they were beautifully made; and the men could handle them skilfully.

There is a big surf running here, and it is a wonder how the islanders can navigate their canvas boats as they do ; but they are experts at it.

We got permission from our Commander to go ashore in watches. So the first batches were put ashore and as soon as we reached it, the islanders started to barter with us for old clothes.

They also wanted tea, sugar, coffee, milk, and any other eatables they could get; while they offered in exchange cured sheep-skins, penguin-skins, knitted socks, and various kinds of curios.

The sheep and cattle here are like skeletons, for the pasture is very poor, and they cannot get hardly anything to eat.

The huts the inhabitants live in are built of huge blocks of stone, which they must have had some trouble to get into position. Even the seats were made of slabs of stone with pieces of wood from some wreck.

The people, apart from the boatmen and those who came forward to deal with us, were very bashful, and seemed at first as if they had no wish to speak to us; but as soon as we spoke to them, they seemed to get interested.

The Settlement, Tritan Da Cunha

Descendants of William Glass, Founder of the settlement on Tritan Da Cunha

John Glass and family

The Mission House on Tritan Da Cunha

Tritan women twisting wool

The Tritan method of grading wool

While we were on the island a new minister arrived, together with his wife. The latter was to be the new teacher for the children.

It was eighteen months since they had had a ship call at the island, with news from the mainland, and they had run out of stores.

We left them plenty of stores of all kinds, putting these in charge of the minister to share out.

The only names I heard were Green and Jones, and the original bearers of them are supposed to have been ex-naval men.

Henry Green's Cottage on Tritan Da Cunha
The flag seen top right was presented by the British Government to the islanders for bravery in saving lives from shipwrecks

The oldest inhababitant on Tritan Da Cunha, Miss Betty Cotton (Aged 95 years)

The island was infested with rats, which eat or damaged everything they can get at.

Before we left this island we collected all the mail and parcels, to take to Cape Town with us. We have left Tristan da Cunha and are on our way to Cape Town; and we are in for bad weather.

To-day it has started to blow a hurricane and the rain is coming down on torrents. I have never seen it rain so heavily or seen such big rain drops. The ship is rolling very heavily, and the Black Squad below knows all about it.

We are continually cleaning the bilge from the pieces of coal and small ashes which keep collecting in them from the sides of the boilers; and we are always getting seas down the engine-room and stoke-hole.

At last our bilge-pumps have given out, and we have been obliged to rig up a hand-pump to try to keep the water down. We are taking this work in spells of half an hour each, till we can get the repairs done.

We are still taking soundings every two or three hours.

To-day the weather has calmed down a little, and we are going to try to get cleaned up as much as possible while we are getting fine weather, as we shall not have very much time for that when we get to Cape Town.

We are cleaning the paint-work in the engine-room, ready for painting before we get our visitors aboard.

We expect to get to Cape Town in two or three days, if the weather holds good.

After three days' fine weather, we have finally reached Cape Town; but we have to lie outside till we receive orders to go alongside the quay.

There are a lot of motor-launches going round the ship, but we have been ordered to keep everyone off the vessel till we are given different instructions.

Our Commander has now gone ashore for his orders, and we shall most likely be going alongside when he returns.

The launches seem to be good business in carrying people round the ship. There are about 20 to 30 of them.

We received orders to go alongside at 3 p.m. and then all were busy getting ready to moor ship.

We were alongside at 3-30, and there were thousands of people lining the quay, waiting to see the little Wonder Ship, as everybody had read about her in the papers.

After the ship was properly moored, and cleaned up, the officials came on board.

Then Mr. Rowett, our head financier, came on board, and praised up the expedition; and after that the ship was thrown open to the public.

We have only allowed 50 people at a time on board.

Everybody wanted a keepsake, and we gave everyone a small souvenir till we had no more to give; and then we started autographing till our fingers were tired with writing. We also had our photographs taken.

We had a large number of reporters on board, all of whom were eager for news of the expedition, but we were banned from giving them any information concerning it. So they had to satisfy themselves by taking photographs.

There were always cars on the quay at our disposal, and their drivers were always ready to take us wherever we wanted to go. These cars were put at our disposal by those responsible for the expedition, and they were a great help, as we were a good way from the town, we had a visit from General Smuts and his daughter, and also from some of the well-known families of Cape Town.

Mr. John Quiller Rowett

There were plenty of invitations, and it took us all our time to fulfil them.

We are still having plenty of visitors. To-day all the Hospital Staff or as many as could be spared were on board, and they were a jolly party.

But it was a great job showing them round, and answering all the different questions they put, and as soon as one batch went, another came.

We had our gramophone playing-or rather the visitors had most of the time.

The following day we were invited to go sightseeing to Table Bay, and that turned out to be a splendid holiday. I got some silver leaves which grow there, for a souvenir when and I had had "*Quest*" R.Y.S., painted on them they looked fine.

At night we want by invitation to the theatre, and that, too, turned out a success. Afterwards we had a party and then turned our way homewards to our hotel, for bed, after a good day's enjoyment.

And what a night's rest we had, after being tossed about all these months, without being able to get a good night's rest! And then to finding one's self laying on a good feather bed! I can tell my readers that I never enjoyed my sleep as I did that night!

We got orders the next morning that the Prince was coming to inspect the ship, and so all was excitement, and we were all kept busy cleaning ship.

The Prince came on board at 1 p.m., steaming alongside in his steam-pinnace, flying the Ensign.

We lined up, and as the Prince stepped aboard, our' Commander saluted. The Prince then made a speech, and then after shaking hands, departed with a good cheering from the public.

We are staying at Cape Town for two weeks, with the ship open to the public all the time. Meanwhile we are having a lot of repairs done, as we sustained much damage to our decks, from the ice.

After we have done our repairs, we are going to Simon's Town, which is only two or three hours' steam from Cape Town.

We left Cape Town for Simon's Town, and as soon as we arrived at the latter place, we received an invitation from H.M.S. *"Lowestoft"* to Dinner, and also to the Cinema Show, which they had on board. We had a splendid time on the *"Lowestoft"*, - the boys could not do enough for us.

To-day we are going to adjust our compasses, before we leave for our homeward trip.

Mr. Rowett came on board and wished every body good-luck and a safe return to then' homes and dear ones.

We left Simon's Town amidst the cheering of the public and the blowing of the sirens of the ships in the harbour.

'We had a splendid send-off.

We were soon out in the open water again, on our last lap for the dear old homeland. But we are calling at Ascension Island en route.

Presently a big passenger steamer steamed close up to us, and when she got abreast of us, all the passengers gave us a good cheering. Mr. Rowlett, our Director, was on board and he gave a blast on the whistle for a farewell.

Commander Wild's Watch - McIlroy, Carr, Wild, Macklin

*The "Black" Watch - **Ross**, Angles, Young, Karr, Smith*

Commander Worsley taking observations of the sun by sextant

Hussey (taking sea temperatures) Commander Wild and McIlroy

CHAPTER VI.
At Ascension.

AFTER a lot of buffeting we arrived of Ascension, and looked around for suitable anchorage. At last a launch came off and showed us a good place at which to drop anchor.

This isle is a central station for the Cable Company; and it is also a Naval Base.

After clearing up the decks, we were told we could go ashore. We found the island a splendid place, but very warm and we were now always wishing for cooler weather. We were, indeed, feeling the heat very much, such a hot climate, after the intense cold we had experienced, proving most uncomfortable.

As soon as we landed, we were given fresh milk and fresh fruit. We had plenty of fruit on board, but it had all been tinned, and the fresh fruit was a welcome change.

We saw a fine lot of bungalows all spotlessly clean, and with everything seeming to be in its place.

We visited the ponds where the turtles are kept and what a wonderful sight they were! Big ones and small ones, and others no bigger than a babies hand! They made a splendid picture; and did not seem to be afraid of us.

These turtles are owned by the British Government, in and are shipped away by them whenever they are required.

The waters round here are infested with sharks, and the Naval Boys catch them, and make fine keepsakes out of their jaw-bones.

We are going to bunker here, and then call at the Azores on our way to England. After a bad trip, we got to the Azores, and spent a day doing a lot of small repairs and bunkering.

We are now proceeding to sea for our last lap to England. We are going to Plymouth, after getting a good wash down, we reached Plymouth, where our Director, Mr. Rowett, met us, and addressed us. We then got our orders to pack up our kits and get them to the station, where we are bound for London.

We have finished our expedition here, at Plymouth, but we are going to London for the settling up and a farewell party, after which we shall go to our different homes.

This expedition was my first experience of the ice, but I hope it will not be my last one. I think indeed, I must have got the Ice Fever, for having been on one such trip, I am anxious to go on another; and I hope and trust it will not be long before I succeed in doing so.

THE SHIP'S COMPANY

MAPS

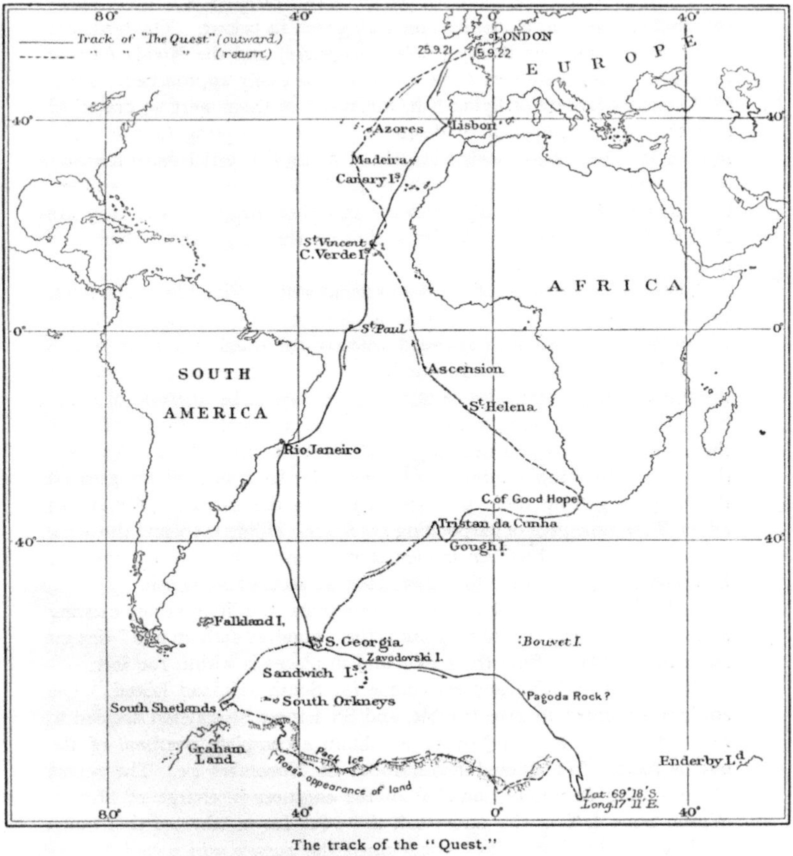

The track of the "Quest."

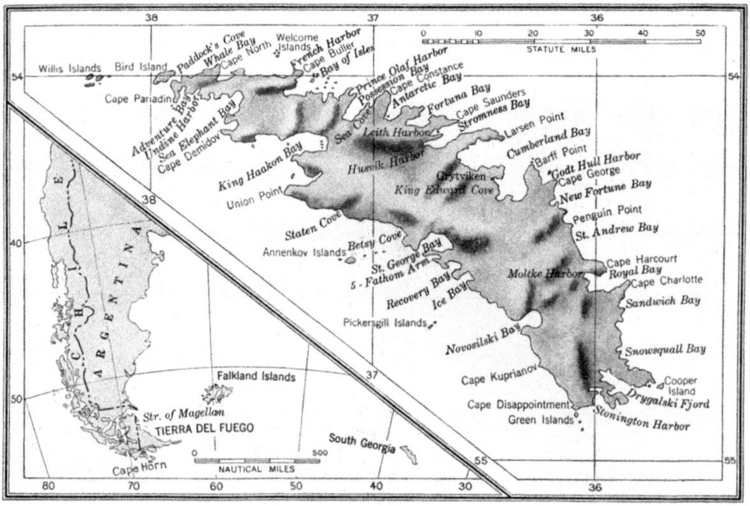

THE ISLAND OF SOUTH GEORGIA, OUTPOST OF THE ANTARCTIC

It lies in the South Atlantic, more than 1,000 miles east of Cape Horn and 2,500 miles south-west of the Cape of Good Hope.

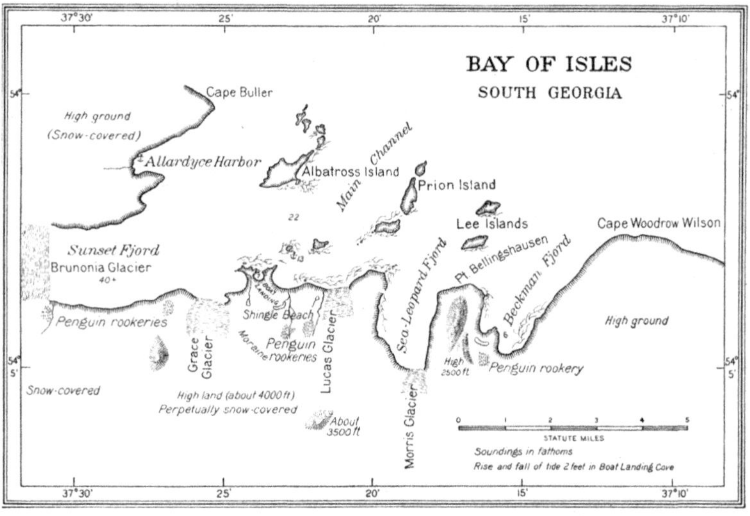

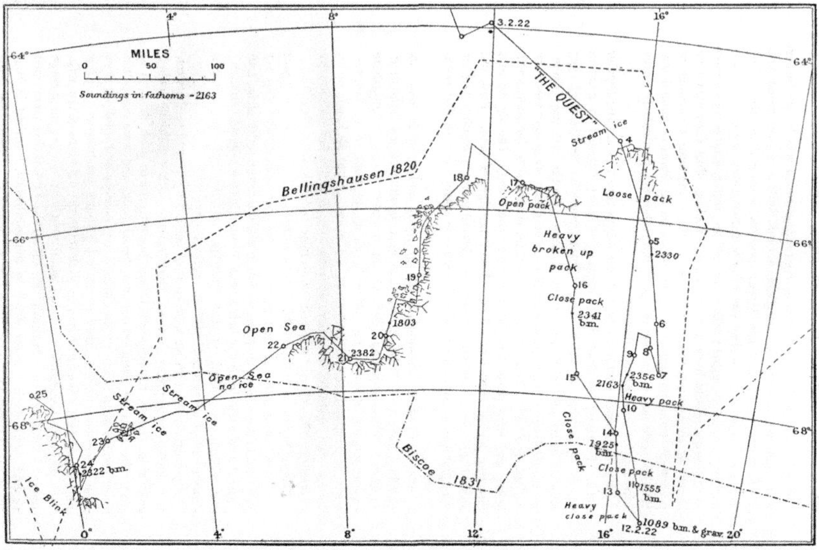

Track of the "Quest" in the far south, 3—25 February 1922.

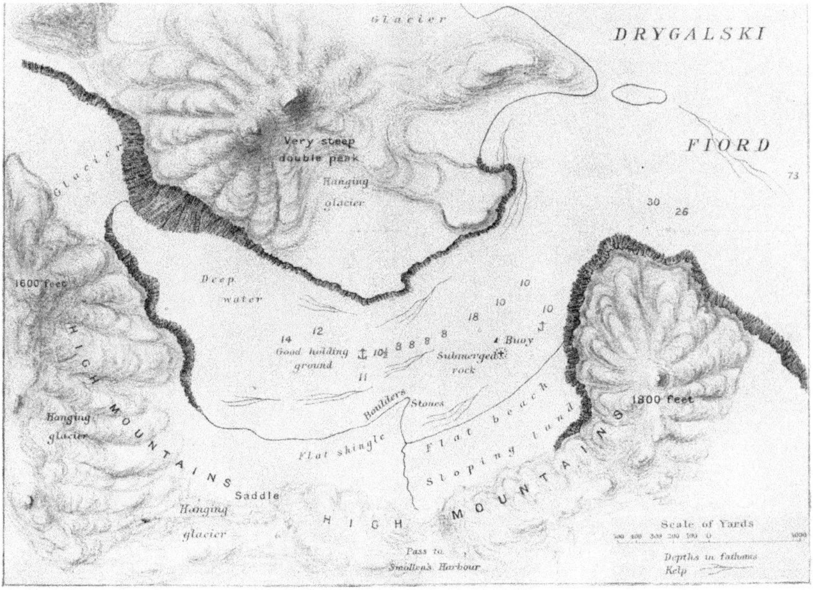

LARSEN HARBOUR, SOUTH GEORGIA

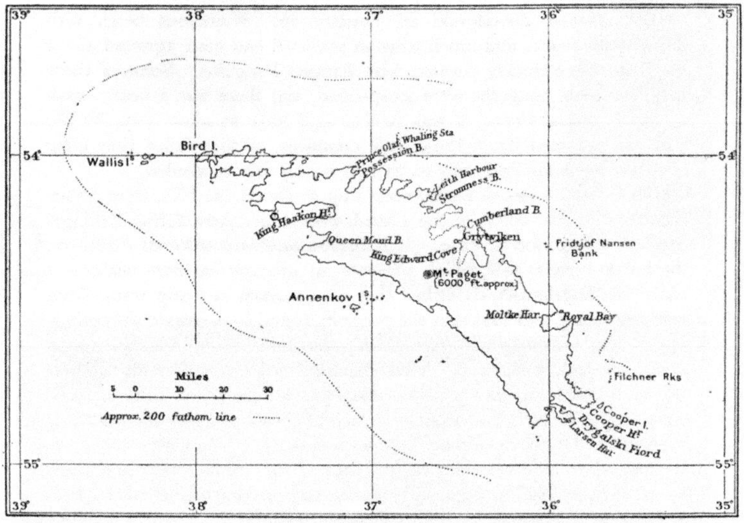

South Georgia.

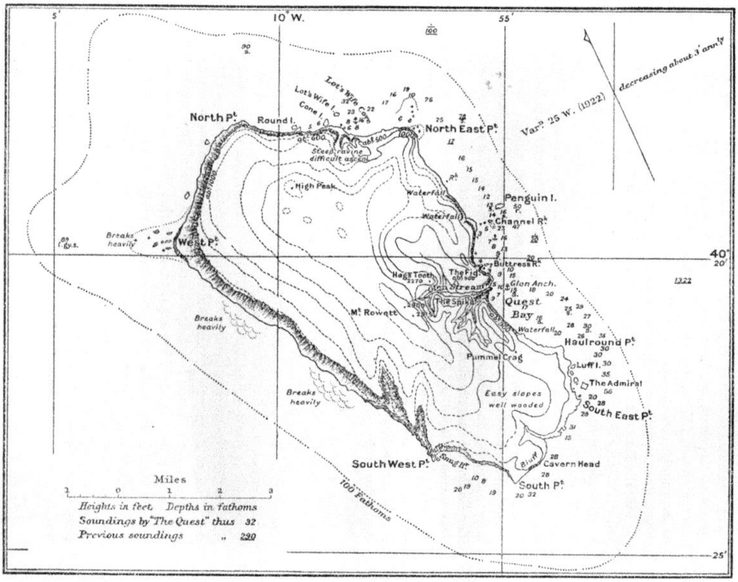

Gough Island.

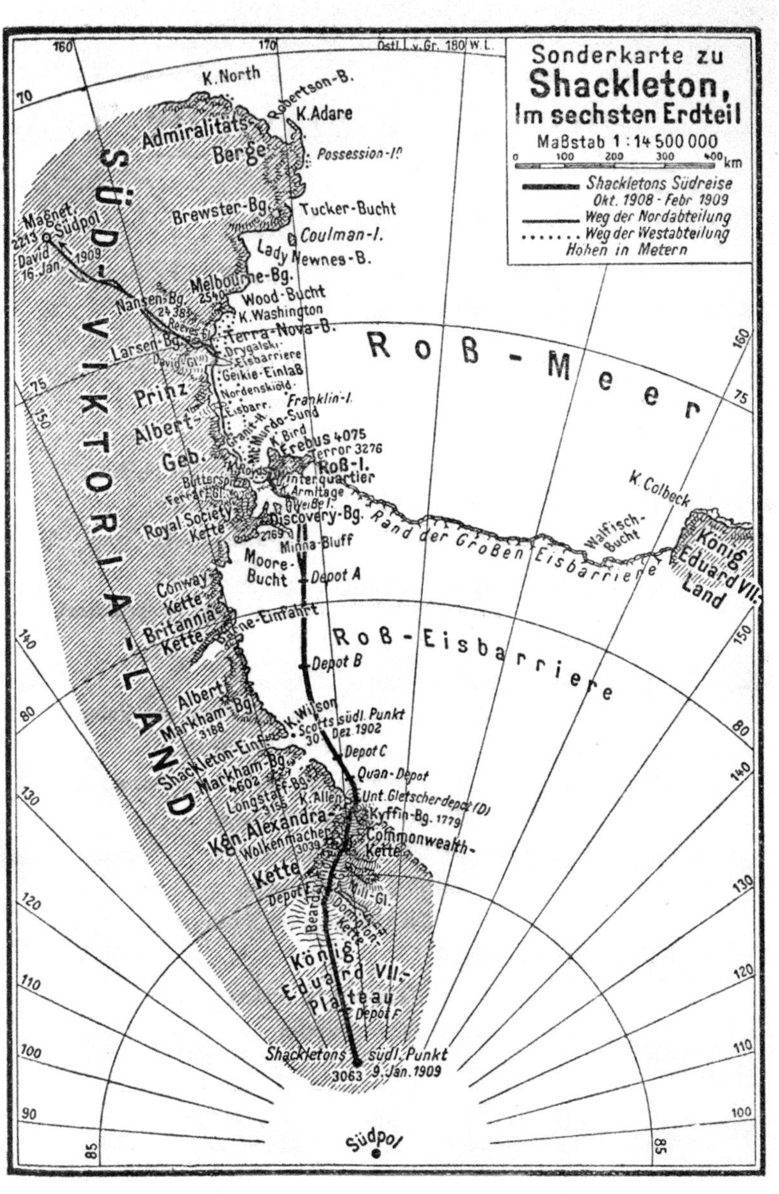
Sonderkarte zu
Shackleton,
Im sechsten Erdteil
Maßstab 1 : 14 500 000
Shackletons Südreise Okt. 1908 - Febr 1909
Weg der Nordabteilung
Weg der Westabteilung
Höhen in Metern
Süd-Viktoria-Land
Roß-Meer
Roß-Eisbarriere
Rand der Großen Eisbarriere
König Eduard VII.-Land
König Eduard VII.-Plateau
Shackletons südl. Punkt 9. Jan. 1909
Südpol

The Quest Chronicle the Story of the Shackleton-Rowett Expedition by Jan Chojecki, grandson of John Quiller Rowett

In 1921, a converted Norwegian sealer, *Quest*, carried the legendary explorer Sir Ernest Shackleton towards the Antarctic on what turned out to be his last expedition. Shackleton had acquired *Quest* with the aim of exploring the Beaufort Sea area of the Arctic.

In 1920, when Shackleton was planning the expedition, his family were living in Eastbourne in a house provided rent-free by the London businessman John Quiller Rowett. Rowett, who had made a fortune in the import of rum, had been an acquaintance of Shackleton when they were at school together at Dulwich College. Rowett provided some cornerstone financing for Shackleton's Arctic project. The understanding was that the majority of funds should come from the Canadian government. In return, Shackleton gave Rowett the *James Caird*, the small boat in which Shackleton had made the remarkable escape from Elephant Island to bring rescue to the stranded crew of Endurance. In late May 1921, when Shackleton finally accepted that Canadian funds were not going to be forthcoming, John Quiller Rowett agreed to finance the entire venture. By this time, it was too late to conduct an Arctic programme that year and Shackleton turned to an alternative plan – a circumnavigation of the Antarctic and exploration of lesser-known waters and islands of the southern oceans. The plan was to depart in late summer, which gave little time to switch plans and prepare for a completely different set of objectives and itinerary. Rowett was a thoughtful benefactor of modern science and was insistent that *Ques*t should execute a programme of research and exploration with topical and worthwhile aims for social and economic impact. Consequently, *Quest* was packed with technology and scientific equipment, though not the trained staff to operate it.

Quest departed in September 1921, heading south through the Atlantic. Almost immediately it became clear that the vessel had numerous problems. The engine was underpowered and had mechanical problems, the sails and rigging insufficient, the stability extremely poor. Progress was slow and uncomfortable – intolerably so for some, who had to be sent home. Shackleton decided to head to Rio de Janeiro instead of Cape Town, with the consequence that the expedition never obtained the key supplies, personnel, and equipment to have been picked up there. At Rio major repairs were conducted while the crew languished ashore. The expedition's naturalist, George Hubert Wilkins, and geologist, George Vibert Douglas, were sent on ahead to South Georgia to commence scientific surveys. Shackleton was highly stressed by the delay and weighed down by the string of society events put on in Rio for the distinguished visitors.

The patched-up *Quest* finally arrived at South Georgia on 4 January 1922. In the small hours of the following morning Shackleton died of a massive heart attack in his cabin, attended by the ship's doctor and quartermaster, Alexander Macklin.

With Shackleton gone, his stunned comrades laboured on under the command of Frank Wild, the Antarctic veteran and trusted right-hand of Shackleton. They flirted with the ice of the Weddell Sea, wary of becoming beset like Shackleton's previous ship *Endurance.* They returned to Elephant Island, for some stirring up memories of their last visit. They visited uninhabited and sparsely populated islands of the South Atlantic, such as Gough Island and Tristan da Cunha.

But back home, with Shackleton leaving massive debts, receivers were appointed to control the expedition. John Rowett's hands were tied. *Quest* was recalled with its mission unfinished and unfulfilled.

Was Rowett's sponsorship a generous gesture of friendship, was it well-intended philanthropy, or was it a business investment? Why did the book of the expedition, *"Shackleton's Last Voyage"*, published in 1923, name Frank Wild as the author when in fact it was written chiefly by Alexander Macklin? Why did the "up and coming" Australian explorer, George Hubert Wilkins, choose to go with Shackleton on *Quest* instead of joining the expedition of Stefansson, the Canadian explorer? And why did Wilkins not put his name to the account of his natural history work on the expedition?

Shackleton's Last Voyage did not tell the whole story. It did not confront the intrinsic problems with the vessel and with the planning that haunted the endeavour throughout, the difficulties arising from Shackleton's untimely death and the harsh realities of the break-up of the team upon *Quest's* return that handicapped dissemination of the outputs.

Despite the altered and curtailed itinerary, the expedition nonetheless did carry out a considerable volume of research. Wilkins and Douglas managed to spend several weeks studying on South Georgia. The visits to the various south Atlantic islands were also productive.

J Chojecki, Scouts and JQR on Quest departure

A large collection of biological specimens made its way in various shipments back to the British Museum of Natural History. Its breadth of taxa and geography sets it apart from the collections of other polar expeditions and together these materials are still a key resource for Antarctic and ocean science.

And the Shackleton-Rowett Expedition launched the career of James Marr, the scout who – together with Norman Mooney – was selected to be Shackleton's cabin boy. Marr went on to a distinguished career in polar exploration and science and was instrumental in establishing what became the British Antarctic Survey.

The Quest Chronicle is the first book focussing solely on the Shackleton-Rowett Expedition for almost a century. Compiled and written by John Quiller Rowett's grandson, and including unpublished archive material and items from hitherto private collections, *The Quest Chronicle* reveals, through the actual events and words of those involved, the deeper story of the expedition, its aftermath and its eventual, surviving, legacy.

Jan Chojecki

The Quest Chronicle,
published September 2022:
www.questchronicle.org.uk
Hardback
ISBN 978-1-913719-72-2
Paperback
ISBN 978-1-913719-71-5

Geraldine Wythe

The following section on the book covers photos of Ernest Shackleton and the crew of the *Quest* getting ready to leave on the expedition.

Ernest Shackleton

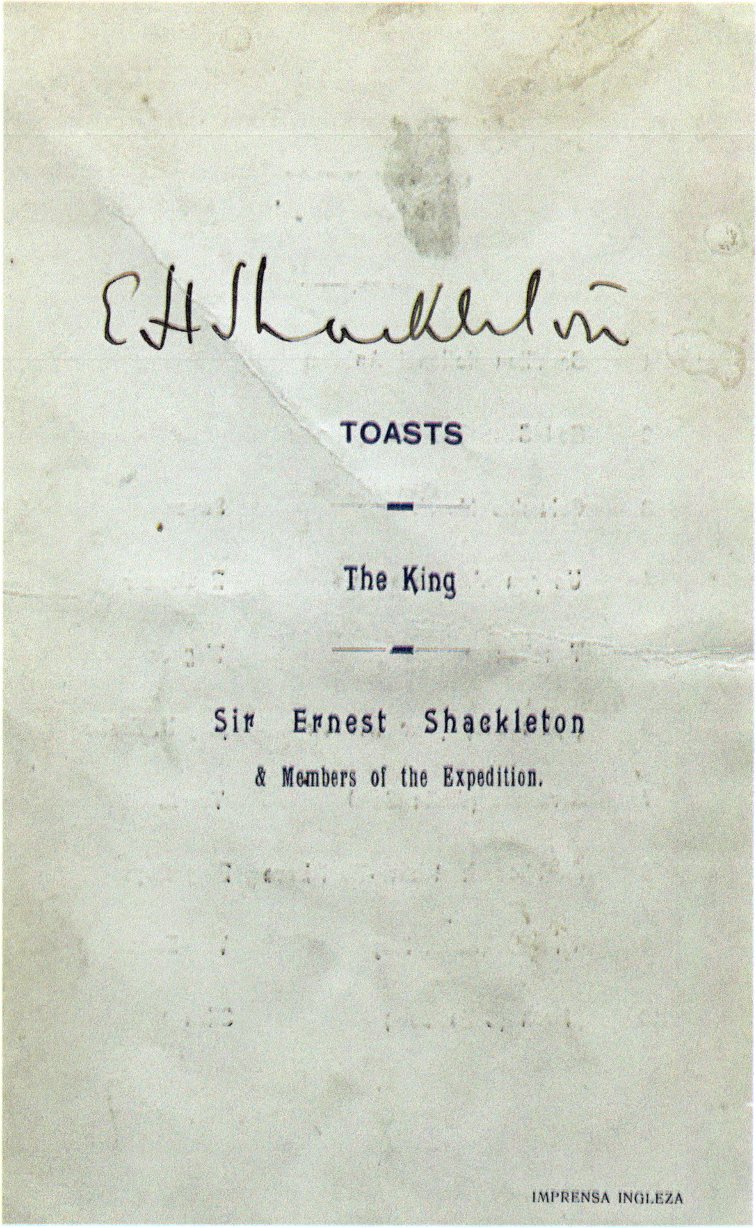

E H Shackleton

TOASTS

The King

Sir Ernest Shackleton

& Members of the Expedition.

IMPRENSA INGLEZA

BRITISH SOCIETY.

RECEPTION

TO

SIR ERNEST SHACKLETON.

and Members of the expedition

CLUB CENTRAL.

28th. NOVEMBER 1921.

RIO DE JANEIRO.

Frank Wild

Dr Alexander Macklin, Alexander Kerr and James Marr.

Sir Ernest Shackleton and John Quiller Rowett

Sir Ernest Shackleton with his son Edward, far left Worsley, far right Wild

Sir Ernest Shackleton and his wife Lady Emily Shackleton

Sir Ernest Shackleton on the Quest

Ernest Shackleton and Wild

The Ward Room on the Quest

Charles Green, the cook with Ernest Shackleton

GREEN MCILROY JEFFREY MARR CARR *F. L. Horton*
MACKLIN Visitor SHACKLETON WILD WORSLEY Visitor
KERR

Alexander Kerr bottom left, the ships dog is called 'Query'.

Crew of the Quest

Ernest Shackleton

Ernest Shackleton

Editors Note:

This pushes the bounds of Journalistic licence when the Telegraph Reporter says Shackleton died in his arms, and he made the cross, in other words it can make a great heading but is very economical with the truth.

Over the years George Ross has mostly been ignored by authors writing on the Quest Expedition, due I expect mostly to what this reporter wrote after his death.

Shackleton died in his arms

MR. GEORGE HENRY ROSS, of 18, Havelock Road, Yarmouth, who was with Shackleton on his 1921-22 expedition to the South Pole, died at Northgate Hospital on Sunday, aged 76.

Sir Ernest Shackleton, he recalled, died in his arms on January 5th, 1922, and he made the cross for the explorer's grave in South Georgia

Mr. Ross was one of the last survivors of this last, ill-fated Shackleton expedition. He was whaling in South Georgia when he joined the ship Quest as a donkeyman.

He was born in London and, at the start of an adventurous career, went to sea when he was 14.

In 1913 he went to the Shetlands, where two years later he married a Shetland girl.

Mr. Ross was one of the first men from the Shetlands to join up at the outbreak of the first world war. He was in the battle of Jutland, and was seriously injured while serving in the Royal Navy.

Mr. GEORGE HENRY ROSS
Shackleton Expedition Survivor Dies
Daily Telegraph Reporter

Mr. GEORGE HENRY ROSS, of 18, Havelock Road, Yarmouth, who was with Shackleton on his 1921-22 expedition to the South Pole, died at Northgate Hospital on Sunday, aged 76.

Sir Ernest Shackleton, he recalled, died in his arms on January 5th, 1922, and he made the cross for the explorer's grave in South Georgia.

Mr. Ross was one of the last survivors of the ill-fated Shackleton expedition. He was whaling in South Georgia when he joined the ship *Quest* as a donkeyman.

He was born in London and, at the start of an adventurous career, went to sea when he was 14.

In 1913 he went to the Shetlands, where two years later he married a Shetland girl.

Mr. Ross was one of the first men from the Shetlands to join up at the outbreak of the first world war. He was in the battle of Jutland, and was seriously injured while serving in the Royal Navy.

Lumberjack: After that war he went to South Georgia. Between the Shackleton expedition and the second world war he worked as a lumberjack in Canada.

Then he joined up again for service in the second war. Two of the ships in which he served as a gunlayer were torpedoed.

For a time during the war he was based in Lowestoft. As well as whaling and lumberjacking, the other jobs Mr. Ross tackled including helping in hospitals and fishing. He claimed that there was not a country in the world he had not visited. He travelled round the world eight times.

Mr. Ross came to Yarmouth from the Shetlands after the last war with the Scottish herring fleet. His wife, Mrs. Annie Ross, worked for more than 40 years as a fisher girl. Mr. and Mrs. Ross went into the boarding-house business at Yarmouth.

They gave up their Albert Square boarding-house last September and moved to Havelock Road. The only thing they insisted on taking with them was the name of the house -- Shackleton House. Mr. Ross also leaves a son. The funeral took place on Wednesday.

Shackleton, his Funeral and the Shetland Pallbearers
by Jon Sandison

The 100th Anniversary of the death of Sir Ernest Shackleton who died suddenly of a heart attack, aged 47, on 5th January 1922 in South Georgia en route to his spiritual home of Antarctica. A faded photograph of the six pallbearers, standing by Shackleton's graveside, which occasionally resurfaces, results in astonishment and surprise that local men took part in such an important ceremony, so far away.

The 5th March 2022, was the 100th Anniversary of the famous explorer's funeral, at Grytviken, South Georgia. It was also a day that six Shetlanders, were selected to be his pallbearers. Upon 12 local hands rested an extraordinary man, so well known the world over, who held a definitive part in our nation's history. That said, the story of the six Shetland pallbearers is not one which is particularly well-known in our local history.

Little wonder perhaps. Shetlanders don't make a big fuss about things, continuing with tasks quietly and steadfastly, a trait that will hopefully continue. But, 100 years on, this part of our island tale should be remembered.

Grytviken Church

Six Shetlanders, Pallbearers.

What perhaps made these men special was not even that they were pallbearers to Sir Ernest Shackleton, but that they - like so many others - had served and returned home to our islands from the experience of the Great War. All of them ex-servicemen who were employed by Leith Harbour Whaling Station, and all from Lerwick: John Byrne, (14 Hangcliff Lane), James Brown (3 Bains Court), Brothers James Leask and Magnus Leask (Upper Lochside), George Manson (14 Burgess Street), and William Sandison (3 Freefield). Also present was Captain Laurence Leask of Gulberwick. Why was it that a group of Shetlanders were selected to undertake such an important and solemn duty at the funeral of one of the most famous explorers of all time?

All the pallbearers had worked within the fishing industry before the war. Fishing in the early twentieth century saw a boom, but one which declined with the onset of war. The war disrupted things with the call up of men.

The Shetland fishing industry began to decline before the war, and this continued after the war with fewer people being employed and the amount and value of the catch variable. A major change was the loss of German and Russian markets for salt herring, with both Russia post

1917 Revolution, and the German Weimar Republic building up their own herring fleets to reduce the dependence on imports. This had a devastating impact on the Shetland salt herring industry. There was little fishing during the war resulting in the stock of herring expanding and good post 1918 catches, but the main market had declined, and continued to do so for many years. Thereupon, following the end the Great War, many ex-servicemen arrived home to a changed local economic and social environment, with scarce employment and a community affected greatly by war, alongside other more personal challenges that faced all ex- servicemen. It was time to look elsewhere for employment and whaling seemed a sensible option. A subsequent lack of employment forced many Shetlanders to move to where the work was.

Accordingly, South Georgia and the Shetland connection is well known through the experience of many Shetlanders at the Whaling. Given some of these six men had been coopers prior to war, this was a trade that was needed at whaling stations. Whaling in the Southern Ocean took hold following the decline in whale stocks in the North Sea around Britain, Iceland and Norway. It had been important to the Shetland economy for over 300 years, beginning with subsistence whaling in the 18th and 19th centuries, and progressing to Arctic and Greenland whaling in the 18th and 19th centuries. In 1909, the British Colonial Office established a whaling base in South Georgia. From there on, whaling expanded in the Antarctic seas. For these men, setting forth to the whaling during the winter months in South Georgia in the South Atlantic, between 1921 and 1922 was an economic necessity, but also perhaps a place of post-war escapism. Whatever the reason, it put these six Shetlanders right at the heart of this historic moment and this extraordinary man.

So what of Ernest Shackleton? The exploits of Shackleton are well known and recorded. But they are always worth repeating; not least one hundred years on from his death. Born in County Kildare Ireland on February 15th, 1874, he was the son of a doctor. His father wished that he follow him into the profession. Being independent of mind, he joined the Merchant Navy and achieved his master mariner's ticket in 1898.

He was a member of Scott's *Discovery* expedition of 1901-04. During the expedition he was responsible for stores and catering on board the ship, and was selected by Scott to be part of a sledging trek to the South Pole.

Southward Bound

(The Antarctic ship *Discovery* pressing southward towards the Pole.)

(*Drawn for the "Boy's Own Paper"* by Charles J. de Lacy.)

Following severe conditions and the onset of scurvy they had to turn back. Further expeditions were separate from Scott, with newspapers in March 1909 reporting that he had made it within ninety seven miles of the South Pole during the *Nimrod* expedition. It was during this expedition that Shackleton obtained his nickname "the Boss" due to his leadership skills. Within all this, stories of spirit, determination and hardship with Shackleton and three comrades just making it back to base alive. A knighthood followed. This was to be the closest anyone would get to the South Pole until the subsequent exploits of Scott and Amundsen in 1912.

ENDURANCE

Then came the *Endurance* expedition. With both Poles conquered, Shackleton came up with a new plan to lead a selected group of men from one side of the southern continent of Antarctica to the other via the South Pole. The story of the *Endurance* and the Imperial Trans-Antarctic Expedition of 1914-1916 is well known. This would present a challenge of covering 1,800 miles in just 100 days. A journey that had never been considered before. Funds, supplies and men were raised for the trip, with money received, for instance, from James Caird, millionaire jute manufacturer from Dundee. Expedition headquarters were established in central London.

Nicholas Reardon

Two ships, the *Endurance* and *Aurora* set forth on the expedition with the main push being undertaken by *Endurance.* She would take Shackleton and his team to the Weddell Sea where they would move off towards the Pole and the eastern side of the Antarctic continent. The *Aurora* was to sail to Cape Evans in McMurdo Sound in the Ross Sea, slipping off a team of men whose task was to move into the interior from that side of the continent, working their way back towards their ship, dropping off supply dumps as they went. Shackleton's men would make use of the supplies as they completed the second half of their proposed journey from Pole back out to sea.

The planned expedition, in line with other expeditions, incorporated an element of patriotic duty.

As their *Endurance* was being loaded in London during the summer of 1914, politicians and nations of Europe were moving towards war. Departing Millwall Docks on the 1st of August, this was the same day that Germany declared war on Russia. At this time Shackleton pulled his crew together on the deck, giving them the forthright option of leaving to enlist in His Majesty's armed forces, if they so wished. As their ship docked in Margate, a few did so. Furthermore, Shackleton sent a telegraph to the Admiralty announcing that his ship, all of its supplies and himself were at the disposal of the nation. Winston Churchill, the then First Lord of the Admiralty, sent back a one-word reply "Proceed". On arrival in Grytviken, South Georgia, news was received that the pack-ice was working further north than usual for that time of year. Despite warnings, on 5th December 1914, the *Endurance* sailed off.

PACK ICE

Endurance entered pack ice over the winter of 1914. By mid-January, 1915, the *Endurance* was lodged hard in the pack ice in the Weddell Sea. Despite trying to break the ice with picks and saws, and an effort to open a channel to clear water, they had to accept the reality of being stuck. There was little that could be done other than to wait out the Antarctic winter. They waited, with the hope that the ice flow would take them northwards. The pack-ice had completely closed around them. She began to break up with the pressure of the ice. Priority was given to salvaging as much as possible of the food and supplies from the ship, with 28 men and sled dogs moving to tents on the ice shelf. The men attempted twice to drag small boats into open water.

Using nails and timber taken from the *Endurance*, the heights of the gunwales were raised on the James Caird and Dudley Docker to provide greater protection from swells. On the 21st of November the *Endurance* slid beneath the ice sinking with her bow first and a raised Union Jack fixed to her mast which the men had secured just days before. Supplies of equipment and food were stowed inside.

They were still camped upon ever decreasing ice. When the ice began to break, the men left left for Elephant Island in three small lifeboats. The fight was now to stay alive. The *James Caird*, the *Dudley Docker* and the *Stancomb Wills*, had been named after three of the expeditions original benefactors. Amongst icy seas a course was set for Elephant Island, or Clarence Island, both around 60 miles away and 800 miles beyond South Georgia.

The 28 men set off on 9th April, with food rations dwindled, and some personal possessions they had saved.

They moved across large seas and treacherous waves, with the constant threat of icebergs, cold, frostbite and diarrhoea...

ELEPHANT ISLAND

On 15th April, seven days after setting out from the pack-ice, all three boats and 28 men made landfall on a rocky beach on Elephant Island. Shackleton knew that the men were now too weakened to take their boats further. The next destination would have to be South Georgia, some 800 miles away across desperate waters of the Drake Passage. He decided that just he and the five strongest men would take the James Caird on the final and greatest leg of the journey back to civilisation. If rough sea overtook the six, there would be little hope for comrades on the rock of Elephant Island. On Elephant Island, and prior to the Shetland pallbearers connection, there had already been a Shetland link to the Shackleton story.

March 1916 began with blizzards and desperately cold temperature. To guard against the Antarctic winter, Shackleton set a strict winter routine. He issued all hands the winter clothing originally designated for the shore parties: two Jaeger wool shirts and combinations, or long underwear, Shetland wool mitts, one Shetland wool jersey and, most importantly, Burberry tunics and trousers.

32, Nightingale Place
Woolwich, S.E.
1st December, 1916.

"Dear Madam,

Sir Ernest Shackleton at present is in New Zealand, but I am acting for him here and was Second in Command of the Expedition.

I am very please to be able to say that the woollen goods you supplied to us were excellent in every way, warm, comfortable, wore well and were easily washed.

You may make any use you like of this letter.
Yours truly". It is
Frank Wild.

Mrs A.H. Pottinger
North Nesting,
Shetland.

Shetland Museum and Archives Ref: D1/259/2

The Shetland Archives holds an original letter written by Frank Wild in 1916 who remained on Elephant Island and was in charge of the group waiting for rescue. On his return, he sent it to the businesswoman in Nesting who knitted their underwear made from Shetland wool which essentially helped keep them alive at this time.

Anne Pottinger was the wife of the North Nesting school master and mother of 10 children. She also ran a knitwear business from her home at the schoolhouse, supplying knitwear to the Catfirth airbase crews during World War One. This was a reminder that Shetland was a go-to place for woollen underwear for explorers at the time. However, Elephant island being a considerable distance from shipping routes forced Shackleton to make an unenviable decision. In May, 1916 he, and five other men, embarked on an 800 mile voyage, across grim seas, on the James Caird a 22-foot boat before reaching South Georgia, with the hope of getting further help from the Stromness whaling station. The achievement of Tom Crean, Frank Worsley and Ernest Shackleton has arguably never been equalled, nor surpassed. Having arrived early on 19th May, 1916, their challenge was not over of course given they landed on the southern side of South Georgia, with fellow humanity being on the north side.

With stamina, fortitude and perseverance, that is forever hard to comprehend today Shackleton, with Crean and Worsley decided to take the final journey overland. Depending upon the survival of these three men, were three further men left behind in a cave in Haakon Bay, and the further 22 men stranded beneath two upturned boats on the desolate of Elephant Island, 800 miles to the south across the Southern Ocean. They eventually arrived at the whaling station at Stromness, on the north-east coast, and - at last - fellow humanity. This, a distance of around 40 miles, but they were going across unchartered territory. Upon arriving at Stromness and getting help Shackleton apparently said:- “We have been away so long. Tell us about the war. When did it end”. The great British offensive on the Somme on the Western Front, was just 42 days from commencing.

The men were already both mentally and physically exhausted from months of endurance in the toughest environments on earth. Not long after, the three crew members were rescued from the other side of South Georgia, whilst the remainder of the crew were rescued from Elephant Island later in the year on the 30th of August, 1916, after 105 days. The Ross Sea Party who had been laying depots for Shackleton on the other side of the continent were relieved in October, 1916. Following all of this, Shackleton returned to Britain, which was in the depths of the First World War. He saw service in South America and Russia. Many of his crew from *Endurance* who returned to Britain, went on to see service during the Great War, some losing their lives or being wounded.

THE LAST JOURNEY SOUTH

With war over, by 1920 Shackleton once again turned his attention to the Antarctic with an aspiration to return to the polar region. Money was raised and an old Norwegian sealer ship was renamed *Quest* at the request of Emily Shackleton. They departed from London on September 17th, 1921, seen off by a large, cheering crowd then leaving Rio de Janeiro on 17th December. Frank Worsley was to be captain again, whilst Frank Wild was Shackleton's deputy and Alexander Macklin one of the three doctors on board. The key aims of the expedition were oceanography, looking for sub-Antarctic islands, mapping unchartered coast and looking for Captain Kidd's treasure in Trinidad and for pearls in the Pacific. In Rio Shackleton suffered a heart attack but refused to be examined, let along turn back. He recovered and the *Quest* continued south. Come 4th January, they sighted South Georgia after a stormy passage. An anchor was cast at the Grytviken Whaling Station. There, Shackleton found a number of old-stagers still manning the station at Grytviken. He was warmly greeted by Fridthjof Jacobsen, still Grytviken’s manager, and the men went ashore and looked after old haunts where they had passed a month while *Endurance* laid at anchor.

At the end of the day they returned to the ship and had dinner on board.

DEATH

Shackleton went on shore to visit old associates and make preparations for getting his ship ready to proceed on the much-reduced Antarctic cruise that the later date permitted. In the evening he returned on board, apparently quite well. He told his comrades that the was happy and contented and feeling that the voyage had really begun. He died later at 3:50am on 5th January in his cabin from a massive heart attack which took him. Alexander Macklin had sat up with him some minutes when he had taken unwell, and suggested that he took the opportunity to take things easy in the future.

"You are always wanting me to give up something" replied the Shackleton. "What do you want me to give up now"? These were apparently his last words spoken. A keen reader of poetry, in his diary his last written words were equally prophetic: "In the darkening twilight. I saw a lone star hover gem-like above the bay".

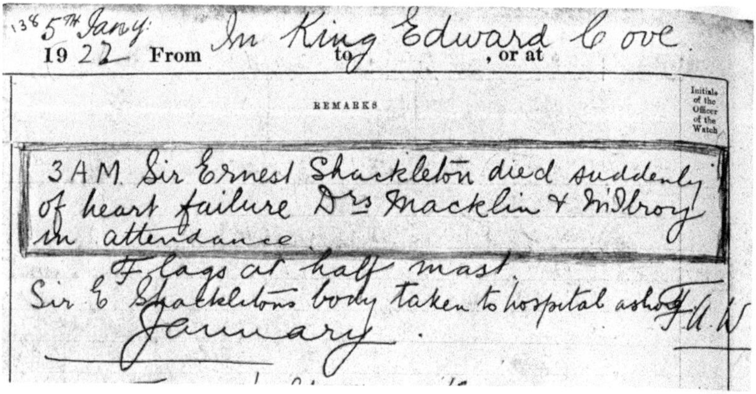

138 5th Jany. 1922 From In King Edward Cove to , or at

REMARKS

Initials of the Officer of the Watch

3 AM. Sir Ernest Shackleton died suddenly of heart failure Drs Macklin & McIlroy in attendance

Flags at half mast

Sir E Shackleton's body taken to hospital ashore F.W.

January.

Deck-log of the *Quest*; Worsley's entry

His body was embalmed and removed to the little Lutheran church as it was decided to send it home for burial unless other instructions were received at first port.

MONTEVIDEO - URUGUAY

Shackleton's closest friend and second-in-command, Frank Wild, made arrangements to send the body home to Lady Shackleton. It was wisely decided that Lady Shackleton should be consulted before the step was taken. It was agreed that one trusted friend should accompany the body back to Britain and see the desired arrangements carried out. The choice fell on Leonard Duncan Albert Hussey, taken on as a meteorologist with Shackleton.

On 19th January his remains were put on the Norwegian steamer *Professor Gruvel* and the long funeral procession began. The body was transported as far as Montevideo, a journey of 10 days.

Whilst his remains lay in Montevideo the Uruguayan Government made the occasion one of national mourning and, in turn, accorded to Shackleton full honours due to a Minister of State. A hundred marines were marched down to the ship's side to bear the coffin draped in a British flag, to the military hospital, where it lay in state under guard.

Montevideo

Montevideo

Val Kerr

Montevideo

THIS SURFACE IS ADHESIVE.

OUR EMPIRE

A SERIES OF 50

44

GRYTVIKEN, SOUTH GEORGIA

Falkland Islands and their dependencies which form Crown Colony are among smallest and most remote British possessions. Falklands are in South Atlantic, 300 miles east of Magellan Straits. Area of 4,600 square miles and population of 2,500. South Georgia is 800 miles to east of Falklands. Area 1,000 square miles. Population 650 men and 4 women. This island is mass of high mountains covered with snow where slopes are not too precipitous. Sir Ernest Shackleton, Antarctic explorer, was buried here. Other dependencies in group are South Orkneys, Sandwich group and Graham's Land.

STEPHEN MITCHELL & SON

ISSUED BY THE IMPERIAL TOBACCO CO. (OF GREAT BRITAIN & IRELAND), LTD.

It was here that Hussey received a telegram informing him of Lady Shackleton's request that her husband be buried on South Georgia, situated on the edge of his greatest exploits. She sent word that, at Shackleton's own earlier request, he was to be buried as far south as possible.

In Montevideo the body lay in state for two weeks and a memorial service was held at Holy Trinity Church. On 15th February, after a ceremony in the English Church attended by the President and members of the Government, alongside a diplomatic body and deputations from the British communities in Uruguay and Argentina, Shackleton's remains were taken through streets lined with national troops to the British ship *Woodville*, bound again for Grytviken, South Georgia. The Master of this ship conveying Shackleton's body was a Shetlander, Captain Laurence Leask, of Gulberwick. He too would attend the funeral.

In Montevideo several bronze wreaths had been cast and then placed on the coffin. Leonard Hussey later brought these to South Georgia to be laid on the grave. The wreaths were from his majesty King George V and the British people, the Uruguayan Republic and the British residents in Uruguay and the French Maritime Society.

The Uruguayan Government had offered to send remains south on one of their war ships. However, this offer was declined given the risk of such a vessel being unable to penetrate the ice which was often encountered round the sub-Antarctic island. Nevertheless, the cruiser Uruguay accompanied the *Woodville* to the limit of territorial waters and fired a salute of nineteen guns in honour of the dead.

THE FUNERAL AND SHETLAND PALLBEARERS

His remains were returned to South Georgia and he was buried in the small whalers' cemetery at Grytviken on 5 March 1922. In all, it was a funeral procession of more than 3000 miles that berthed at Grytviken. The community who lived at whaling stations along the coast was small, mostly Norwegians. However, they too had taken to Shackleton. The funeral service was half in Norwegian, half in English. It was held in the Lutheran church at Grytviken. This was a Norwegian Church, which had been prefabricated in its home country, then built by whalers in 1913. As well as the funeral of Shackleton, there has been several marriages and a few baptisms within its walls. At different times since, the church was used as a library, cinema and store.

Grytviken Church
ANNO
1913
Ann Lane

Grytviken Church

Grytviken Church

Pippa Simpson

The historic events were recorded in *The Times*, 4th May 1922, by Captain Leonard Hussey:

“We arrived in the whaler *Woodville* with the body of Sir Ernest Shackleton at South Georgia, ‘the gate of the Antarctic’, on February 26th. We took the body ashore on March 1st. The coffin was taken to the Lutheran Church with floral tributes. At 3pm, on March 5th, the manager of the whaling stations on the island and about 100 men assembled at the church. Mr Edward Binnie, resident magistrate of South Georgia, and the Customs and other officials representing the Falkland Islands Government, were also present. A small ceremony in Grytviken church was presided over by Edward Binnie, the magistrate, and attended by the managers of the five whaling stations and a hundred whalers and sailors who sang their Norwegian funeral hymn.

Just before the service began, Mrs Aaderg, the only women on the island, placed a bunch of freshly gathered flowers on the coffin. Mr Binnie read part of the Burial Service as there is no clergymen on the island. All present repeated the Lord’s Prayer. After the second verse of the Norwegian funeral hymn the coffin was lifted by six Shetlanders from Leith Harbour Whaling Station and was carried over piles of whalebones

Grytviken Church

Rosie Needham

Grytviken Church

Rosie Needham

and many small mountain streams to the little cemetery on the hill where sailors have been buried since 1846. As the cortege left the church, the bell began to toll, and continued till the coffin was lowered into the grave."

One of these Shetlanders, **James Brown,** was given the responsibility to choose the others for duty as pall bearers. The men he selected were all Shetlanders, his friends who were with him there at the time.

Hussey further added that "in front walked two officials with black funeral banners and behind with myself representing Mr J Q Rowett (who was chiefly responsible for financing the *Quest* expedition), Lady Shackleton and children, and the Expedition where Mr Binnie and the station managers carrying wreaths. At the gate of the cemetery the British and Norwegian flags were flying at half mast, and about 100 men, mostly Norwegians, were gathered."

Further detail By Hugh Robert Mill in "The Life of Sir Ernest Shackleton" written in 1923 added that "… the coffin was borne on the shoulders of six stalwart Shetland Islanders, all ex-service men, up the hillside to the grave, followed by the cortege of islanders". It was further recalled that "the managers even from the most distant stations came to attend it, and one of them, with a fine and understanding sympathy, brought with him from a great distance seven ex-service men from a British Ship, Shetlanders all, to act as bearers on the last stage." The coffin, draped in the Union Jack, was carried in procession to the cemetery by six Shetland Island men, preceded by two men carrying black funeral banners in the Norwegian custom.

Hussey further added that, "We reached the grave and the coffin was lowered by the Shetlanders with the head towards the South Shackleton loved so well".

He added that "... a simple wooden cross was erected, on which were placed the wreaths from Mr Rowett, Lady Shackleton, and the Expedition, the remainder of the wreaths being placed on the coffin. Behind the cross rested the magnificent bronze wreath from Uruguay; in front a palm from the French Maritime Society and wreaths from the British Society of Uruguay and Masons".

Frank Wild, his fellow companion from the Endurance Expedition poignantly reflected on this remote, yet historic burial:

"It is just here that he would have chosen to lie, and would have preferred this simple funeral to any procedure carried out with great pomp and ceremony. The graveyard is a simple little place. An ideal resting-place this for the great explorer who felt, more than most men, the glamour of such surroundings."

Symbolically, his was the only grave with its head to the south, rather than the east. *Quest* returned to England on 16th September 1922.

THE SHETLAND PALLBEARERS

Corporal James Brown, known locally as "Sodger" Brown served with 2/Royal Scots. He saw action at Mons in August 1914, being one of the original "Old Contemptibles", those that had served from 1913 and throughout the war being awarded the Mons Star and Bar. He was twice wounded. Following discharge, he also served on *Ambitious.* After the war he settled back in Lerwick, and with other Lerwick boys went to South Georgia. On returning to Lerwick he worked as a stevedore and docker. During World War Two he joined the Royal Engineers, after being turned down for the Home Guard on account of his age. A period was spent on operations at the docks at Le Havre where he witnessed evacuation of troops in 1940, as part of Operation Aerial. James passed away in Swansea, on the 17th of June, 1966. Upon his death, *The Shetland Times* obituary noted that “Soldiering was the late Mr Brown’s life and first love. He always recollected the good times, never the hard. Always smart and soldier-like he was the smartest soldier in his regiment and always chosen for all ceremonial occasions”. At his funeral, floral tributes were sent by the local Royal British Legion.

James Leask originated from Lochside, Lerwick. He worked as a fishworker. In August 1914 he enrolled as a Seaman with the Royal Naval Reserve (Shetland Section) serving on an X-Class Motor Lighter, which was a shallow craft, used for landing, as well as water and fuel tankers. He did not see action but was awarded a 1st Good Conduct Badge and British War Medal.

His brother, **Magnus Leask,** having been employed as a fishworker, enlisted as a seaman with HMS *Thalia* depot ship, which was parent ship for boats of the Auxiliary Patrol, Cromarty and Peterhead'. The Roll of Service states that he served the Royal Naval Reserve (Shetland Section) as a gunner on hired drifter HMS *Moniadhla.* Following the war, Magnus was employed as a crofter. In the 1970s, Ernest Shackleton's son visited Shetland and was presented with a copy of the photograph of the burial by Magnus. He was also awarded the British War Medal. He passed away in Lerwick on 24 March 1987. Their brother, Thomas, served with the 8th Battalion Seaforth Highlanders. He was killed on the Somme, 11th October 1916. Another brother, Robert, served as a Gunner on Oil Tanker *Thermodore,* and was awarded the Distinguished Conduct Medal .

George Manson originating from Hamnavoe, Burra Isle, and resided at 14 Burgess Street. In November 1914, as so many Shetland men did, he had enrolled as a Seaman with the Royal Naval Reserve, (Shetland Section). He demobilised in March, 1919. His Rank at demobilisation was "Gunner". He was awarded the British War Medal and Victory Medal. He passed away on the 13th of April, 1957. He had two brothers, James and Thomas who served in the R.N.R.

John Byrne, enlisted as a seaman with the Royal Naval Reserve [Shetland Section] in 1916, having previously served with the Gordon Highlanders. He served with HMS *Brilliant* shore staff, which was a depot ship for trawlers and patrol boats of the Auxiliary patrol. then later served on

Ambitious, until his discharge in December 1919. After the war he was employed as a baker and the town lamp lighter. He lost two brothers during the war. James, served with the 8th Battalion Gordon Highlanders, and was killed in action at Loos on 25th of September 1915 while Andrew, 7th Seaforth Highlanders, was killed on the Somme on 18th July 1916. A further brother David, served as a Sergeant in the Gordon Highlanders, and entered France with the initial Territorial detachment in December, 1915, and saw action throughout the war. John Byrne died on 16th August 1959, aged 72.

William Sandison, of Freefield, Lerwick, was one of over 200 Shetland Territorials who left our local shores in June 1915. Prior to the outbreak of war, he had been employed as a cooper at Messrs Slaters barrel factory in Lerwick. He saw action with the 7th Battalion Gordon Highlanders on the Western Front throughout 1915-1918, at the Somme, Cambrai and was wounded during the German Offensive of March 1918, obtaining a Military Medal. During this action he was wounded in the leg, and saw further service on the Italian front, looking after prisoners of war. He passed away on 17 September 1969 and very rarely mentioned his role in the funeral of Sir Ernest Shackleton.

However, he would often speak about and sometimes meet up with his old comrades in the Lerwick British Legion, including having the odd pint with "Sodger" Brown. Following the Second World War, William again turned to working in whaling. A passport photograph of him – by now aged 47 – was taken at Leith on 17th October 1945 en route once more to the whaling stations on South Georgia. It was no doubt quite a different experience from when he had done the same job in his early 20s. Like other men who did this, his wife and son would not see him throughout the winter months.

Kay May
Memory of
ERNEST·HENRY
SHACKLETON
EXPLORER
Born
15th Feb. 1874
Entered Life Eternal
5th Jan. 1922

As well as the Shetland pallbearers, another Shetlander was present at the funeral. **Captain Laurence Leask,** originally from Gulberwick, was Master of the cargo ship *Woodville* when it carried the coffin of Sir Ernest Shackleton from Montevideo to South Georgia for burial. She was a cargo ship. In 1914, Laurence had served as a Mate on the *Farraline* which sailed from Leith to Seville, then a First Mate on the *Meriddo*, of Glasgow, sailing from London, France and the Mediterranean. He was awarded the World War One Mercantile Marine Medal and British War Medal. He passed away on the 11th of June, 1937.

Just days before Shackleton's funeral, on 2nd of March, a large memorial service had taken place in St Pauls Cathedral, London. There a huge congregation assembled, including family representatives of the King and Queen, of Queen Alexandra, and the Prince of Wales, of the Admiralty, and of the Royal Geographical Society. Also present were old comrades who had sailed with Shackleton on the *Discovery,* the *Nimrod*, *Endurance* and the *Quest.*

In 1928 a headstone made from Scottish granite was erected on Shackleton's grave and unveiled at a ceremony by the Governor of the Falkland Islands, Sir Arnold Hodson.

The headstone was made in Edinburgh and was carved by Stewart McGlashen.

It was taken to Barry Docks, Wales, by rail and shipped on *Harpon*, a cargo ship that sailed regularly to Grytviken. Shackleton's love of poetry is reflected in the inscription on the back of the headstone based on a poem by Robert Browning. When people visit South Georgia today, Shackleton's graveside is of special interest.

Geraldine Wythe

There is one further twist to this local connected story.

The story goes that a shirt belonging to Shackleton was taken back by Captain Laurence Leask, also present at the burial but not a pallbearer.

This was still in existence in the 1990s, carefully wrapped in tissue, in the loft of a house in Breiwick Road. Tragically, so the story goes, it was dumped when Council workmen cleared the loft during renovations. Despite this, the legacy of Shetland's connection to Sir Ernest Shackleton lives on in more ways than one.

Shackleton's old comrade from the *Endurance* Expedition, Frank Worsley wrote: "I knew that I should never look upon his like again. He was not only a great explorer: he was a great man … by his genius for leadership he had kept us all in good health … by sheer force of his personality he had kept our spirits up … by his magnificent example, he had enabled us to win through …"

A great man, carried to his final resting place, by six, ordinary humble Shetlanders.

Acknowledgements:-

Special Thanks to Kate Canter, Shetland Family History Society.
Thomas Binnie/ Grandson of Edward Beveridge Binnie (C) Thomas Binnie.
Dr Carol Christiansen.
[Shetland Museum and Archives].
John Goodlad.
South Georgia Museum.
Bruce Sandison, Son of Pallbearer William.

References

https://sgmuseum.gs/death-of-shackleton/
H R Mill, The Life of Sir Ernest Shackleton, (Heinemann Ltd: 1923), p.279.
'Shackleton's Burial' Captain Hussey, The Times, 4 May 1922.
The Endurance. Shackleton's Legendary Antarctic Expedition. Caroline Alexander. Seaforth, 2021. p49

B Sandison, Memoirs of A Lerwick Boy, (The Shetland Times Ltd: 2008), p.4.

'The Last Whalers'. Lyndsie Bourgon. https://aeon.co/essays/the-last-whalers-commuting-from-the-north-sea-to-antarctica

The Endurance. Shackleton's Legendary Antarctic Expedition. Caroline Alexander. Seaforth, 2021.

https://www.nls.uk/learning-zone/geography-and-exploration/shackleton-and-wordie/ernest-shackleton/

Ann Lane

Grytviken in winter

Geraldine Wythe

Leonard Duncan Albert Hussey, OBE

Hussey with his famous banjo

When Shackleton died, Hussey was delegated responsibility for making arrangements for the return of his body to England, and then burial at South Georgia, as requested by Lady Shackleton.

Hussey wrote six chapters for inclusion in Frank Wild's official account of the expedition, but they remained unpublished presumably for lack of space. It has been said by some that Lady Emily Shackleton did not want her husband's body coming home to England and him being lauded as a hero, as she wasn't happy about his on and off affair with the American-born actress Rosalind

Chetwynd (known professionally as Rosa Lynd), who died 9th Oct 1922. a few months after hearing of Shackleton's death.

Lady Shackleton therefore cabled Hussey in Montevideo to take her husband's remains to South Georgia for burial, since this was the site of his greatest triumph.

Left: Rosalind Chetwynd (known professionally as Rosa Lynd)
Right: Lady Emily Shackleton

SHACKLETON sir Ernest Henry of 14 Milnthorpe-road **Eastbourne** knight died 5 January 1922 on board the Steam **SI**
Yacht "Quest" at South Georgia Probate **London** 12 May to dame Emily Mary Shackleton widow.
Effects £556 2s. 2d.

229

£556 in 1922 is worth £40,421 today

Images of Dr. L.D.A. Hussey's full-size medal group, miniature medal group (both mounted as he wore them) and the certificate for his Mercantile Marine War Medal and ribbon for the same (dated Oct. 7, 1937).

The medals are housed in a fitted leather-bound carrying case, with a recessed space for the miniatures beneath the inside tray.

The full-size medals shown below:

"ex-Collection of Glenn M. Stein, FRGS, FRCGS".

The Most Excellent Order of the British Empire
(Military/Officer) – unnamed as issued

British War Medal – named
CAPT. L.D.A. HUSSEY.

Mercantile Marine War Medal – named
L.D.A. HUSSEY

Victory Medal – named CAPT. L.D.A. HUSSEY.

Defence Medal 1939-45 – unnamed as issued

War Medal 1939-45 – unnamed as issued, with
Mention in Despatches Oak Leaf (MID)

Polar Medal/Antarctic 1914-16 – named
L. HUSSEY, METEOROLOGIST, "*ENDURANCE*"

OBE (Military): London Gazette, Jan. 1, 1946
(Squadron Leader, RAFVR)

Mention in Despatches (2): London Gazette,
June 1, 1945 and June 14, 1945

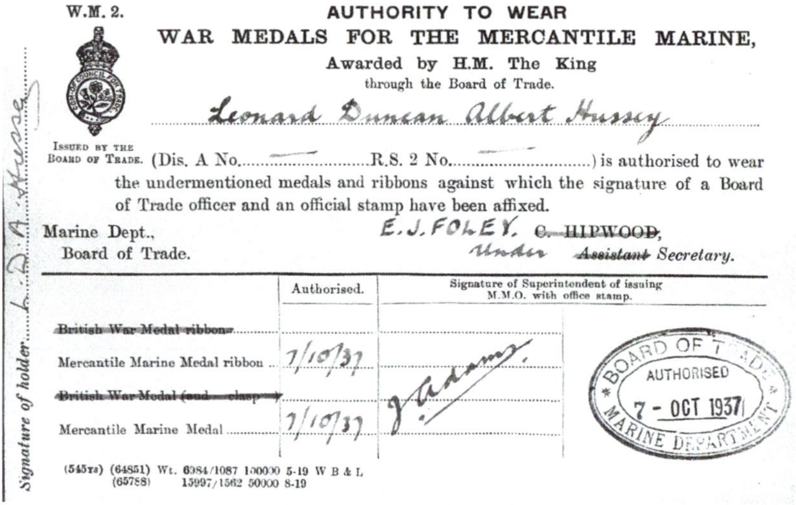

W.M. 2.

AUTHORITY TO WEAR
WAR MEDALS FOR THE MERCANTILE MARINE,
Awarded by H.M. The King
through the Board of Trade.

Leonard Duncan Albert Hussey

ISSUED BY THE BOARD OF TRADE. (Dis. A No.......—.......R.S. 2 No.......—.......) is authorised to wear the undermentioned medals and ribbons against which the signature of a Board of Trade officer and an official stamp have been affixed.

Marine Dept.,
Board of Trade.

E.J. FOLEY. ~~C. HIPWOOD~~,
Under ~~Assistant~~ Secretary.

	Authorised.	Signature of Superintendent of issuing M.M.O. with office stamp.
~~British War Medal ribbon~~		
Mercantile Marine Medal ribbon	7/10/37	J. Adams.
~~British War Medal (and clasp)~~		
Mercantile Marine Medal	7/10/37	

BOARD OF TRADE AUTHORISED 7 - OCT 1937 MARINE DEPARTMENT

(545TS) (64851) Wt. 6984/1087 100000 5-19 W B & L
(65788) 15997/1562 50000 8-19

Signature of holder......L.D.A. Hussey

As they were not officially issued by the government, Hussey acquired the miniatures medals from a jeweller/military tailor and are all unnamed.

There's a nice little story behind the Mercantile Marine War Medal (MMWM). Evidently he gained this award by virtue of having been aboard the *Endurance.* Since she sailed at the very start of the War and was classed as a mercantile vessel, she was at sea for at least six months and certainly made one voyage through a "danger zone"; voyages to all parts of the world from August 14, 1914 – April 11, 1915 qualified for the medal, on account of the German auxiliary cruiser *Kronprinz Wilhelm* and *SMS Dresden*. The former operated off the eastern coast of South America during this period, while the *Dresden* was one of the participants in the Battle of Coronel on Nov. 1, 1914. The battle was a German victory, but Shackleton was unaware of the engagement and certainly did not know that Admiral von Spee was headed round Cape Horn.

Originally, *Endurance* was to sail from Buenos Aires to Port Stanley, Falkland Islands, but at the last moment, Shackleton announced a change of mind. Though he pleaded financial necessity, in reality it was due to fear in the Falklands that von Spee may appear in the waters of the Colony. Shackleton headed for South Georgia Island instead. This decision may have averted disaster.

The *Endurance* left South Georgia on December 5, and just three days later, there was a major naval engagement off the Falkland Islands. This time, a British victory avenged the losses at Cape Coronel.

Now, evidently Hussey wasn't aware of his entitlement until many years later, as he applied for and received the MMWM in 1937.

by Glenn M. Stein, FRGS, FRCGS

Leonard Duncan Albert Hussey and his Dog Team

Leonard Duncan Albert Hussey and Samson

*Raising the Union Jack given by King George V to the **Quest**.*
(Scout Marr is Hoisting the Left Signal Halyard)

Shackleton's Scouts

Norman Erland Mooney & James William Slesser Marr by Alan S. Noake

Scout Rally, St Albans June 1914.

The Hertfordshire Scout County Commissioner, Percy Everett, Leading Sir Ernest Shackleton to the saluting base. They are followed by the Earl of Clarendon, County President and Lord Lieutenant of Hertfordshire and Charles Dymoke Green, District Commissioner of St. Albans.

Shackleton was an avid follower and supporter of Scouting from the outset. In June 1914, a few months before setting off on the Imperial TransAntarctic (*Endurance*) Expedition, and only weeks before the outbreak of war, Shackleton attended the large Scout Rally at St Albans.

At the St Alban's Scout Rally Sir Ernest gave the following Address:

"Addressing the Scouts Sir Ernest Shackleton remarked that he was very much impressed by their charge, in fact he had no idea the Boy Scouts could produce such a result. He had heard Scouts close to his own house; he had heard the drum and bugle going just as the children were being put to bed, and on looking out he had found perhaps two small boys (laughter). But when he thought of what the organisation was, and what it was doing all over the country, he thought it was one of the finest things they had in modern days (applause). He was just a Scout, perhaps in a bigger field, and he knew the interest the Scouts and others took in the work his comrades and himself had to do away in the frozen South." [1]

On 9th July 1921, in the run up to the Shackleton-Rowett '*Quest*' Expedition Shackleton wrote:

"For many years, I have been an admirer of the Boy Scout Movement, which I may say appeals to me particularly because it seems to give every boy a grounding in the practice of exploration."[2]

So, it was no real surprise when Shackleton announced he would be interviewing candidates to take a Scout as cabin boy on the *Quest*. 1700 boys applied and this was whittled down to 10 who met with Shackleton. Eventually Shackleton chose two Scottish Scouts to go – Norman Erland Mooney from Orkney and James William Slesser Marr from Aberdeen.

So why take Scouts? Well firstly, Shackleton already knew the benefits of youth involvement and what they could bring to a project. Secondly, links had been building up for several years prior to the *Quest* expedition between Shackleton and Baden-Powell such as Shackleton attending the large Scout rally. Thirdly, was the influence of Emily Shackleton who went on to play a major role in the development of the Scouts and the Girl Guides. It seems it was her idea in the first place. Fourthly, and most importantly Shackleton instinctively knew (and was proved right) that the press would latch on to the idea of including a Boy Scout in the crew as cabin boy. It generated extra publicity in the Daily Mail for months on end.

Sir Ernest Shackleton (centre) with Scouts Norman Mooney (left) and James Marr (right), prior to the Quest's departure from London, 1921

Front L to R: Percy Everett Deputy Chief Scout, Ernest Shackleton, Major Wade Secretary of the Scout Association.

August 1921 London, Scout HQ.

The timing was good for Shackleton because after WWI there was a huge upsurge in the popularity of Scouting and the first World Scout Jamboree in Olympia had just been recently held in 1920.

The Scout Heritage archives contain two original pieces of correspondence between Shackleton and Baden-Powell. The first is dated 11th August 1921 and the second 10th September 1921. The second was just a week before the *Quest* sailed from London and just shows how quickly the selection of the two Scouts was arranged.

In his Shackleton autobiography (Smith, 2014) talks about the selection of Shackleton's Scouts as follows:

"As a publicity stunt, the idea aroused much attention from the press and public. For this, Shackleton could thank his wife. Since 1917, Emily had thrown a large amount of her time into the Girl Guides and Boy Scouts. With the children at school it was a means of usefully filling her time. Baden-Powell later invited her to take control of the Guides in Eastbourne and, as her interest developed, Emily encouraged scout companies around the world to adopt the Shackleton name."

Mooney really suffered from sea sickness, and it was proposed that he leave the ship at Lisbon, but he stayed on board on the next leg all the way to Madeira. However, upon reaching Funchal, Madeira, Shackleton decided to send Mooney back home to the UK. Marr then had the responsibility of continuing alone as the sole representative of Scouting. He became a strong member of the crew and was present in South Georgia when Shackleton passed away. Following the expedition in 1923 Marr published a full account of his adventures on board the *Quest* in his book, Into the Frozen South.

So, what an exceptional legacy Shackleton's two Scouts have left behind them for their fellow Scouts. Marr and Mooney were selected from thousands to represent Scouting. Mooney's story is very much a real-life lesson in having to deal with extreme disappointment and setbacks. By contrast, Marr's story is one of making the most of every lucky opportunity. Marr had a golden ticket to visit Antarctica as a teenager on *Quest*, then returned several times as a marine biologist, served there during World War Two and ultimately played a very significant part in setting up today's British Antarctic Survey.

A more comprehensive account of the lives of Marr and Mooney is given in the book Shackleton's Scouts by Alan S. Noake.

Publication Sources

1. Hemel Hempstead Gazette 20th June 1914.
2. Daily Mail 9th July 1921.
3. Smith, M. 2014. Shackleton: By Endurance We Conquer. London: One World.

Image Sources:

Scout Rally, St Albans June 1914. [Source: Unknown.]

Sir Ernest Shackleton with Scouts Norman Mooney and James Marr.

[Scout Association Archives – © The Scouts Heritage Service.]

Speeches at the unveiling of the Shackleton Memorial

At the invitation of' the Memorial Committee the company assembled in the Hall of the Society at 3 p.m. on Saturday 9 January 1932.

The Most Honourable the MARQUESS OF ZETLAND, Chairman of the Memorial Committee, addressing the President of the Society, said: My first duty is to thank you, Sir, and through you the Council of the Royal Geographical Society, for agreeing to take into you're keeping the statue which forms part of the Memorial which the Committee, formed for the purpose, have raised to the late Sir Ernest Shackleton.

Of the man himself I shall have an opportunity of speaking later; for the moment I confine myself to a brief statement of the manner in which the Committee has discharged the task entrusted to it. In all a sum of rather more than £3,300, including interest which has from time to time accrued has been at our disposal. Some part of this we were able to allot in the form of an annuity to Mrs. Shackleton, the mother of the explorer, until the date of her death. It had been in our mind to assist, had this been necessary, with the education of Shackleton's children; but at the specific request of Lady Shackleton no part of our fund was used for that purpose.

Part was spent in erecting in South Georgia, in the vicinity of Shackleton's last resting-place, a granite monument; and part will be spent in the acquisition of a bronze bust to be entrusted to the keeping of one of the public institutions in this country. But the greater part has been devoted to the striking statue of Shackleton in Polar costume so finely executed by Mr. Charles Jagger, which is to be unveiled to-day.

Before asking Dr. Mill to address you, it would, I know, be the wish of the Memorial Committee that I should take this opportunity of expressing to Mr. Wordie, a colleague and admirer of Shackleton, our unstilted gratitude for the work which he has done as secretary to the committee. The greater part of the burden has fallen upon his shoulders, and it is not too much to say that without his devoted services the goal which the Committee set before itself would never have been reached. I now call upon Dr. Mill, a member of the Committee, to address you.

DR. HUGH ROBERT MILL:

Twenty-three years ago to-day, on 9 January 1909, Shackleton stood on the summit of the ice-cap 400 miles nearer the heart of the Antarctic than any predecessor, though still 100 miles from his hoped-for goal. Weary, hungry, cold, and disappointed he yet tingled for triumph in the greatness of his achievement. To-day, ten years after his death in the *Quest*, the friends who followed, admired, and loved him meet to enshrine in this temple of Geography a memorial which will mark for ever the high precedence he won amongst the heroes of exploration. In this time of gloom and disillusion it is good indeed to Warm our hearts in the glow of romantic aspiration, high attainment, and happy optimism which his memory can never cease to radiate.

Shackleton trained as a polar explorer on the National Antarctic Expedition of 1901, and accompanied his leader to the farthest south, an advance of 200 miles. He could go no farther then; but in his mind he registered the determination to return and to push on to the Pole. So in 1907 Shackleton brought the immense driving power of his impetuous but persistent energy to bear on the creation of an expedition which should be not only new but of a new kind.

He planned and organized it on new lines; he raised the funds by new means, and to its service he brought new devices of stowage, transport, and equipment. Those novelties of a quarter of a century since were so effective and so essential that they have become the common form of all recent expeditions, and only those who remember the tyranny of the old traditions of polar exploration can understand how great were the reforms made by Shackleton.

He chose a land party distinct from the ship's company, treating all its members on equality, though entrusting each with personal responsibility for a special department of investigation or service. Although he consulted authorities on every point Shackleton did not follow any master; advice or warning only stimulated his originality. He was the first to bring a motor car to the Antarctic, and though its life was short it was a merry and useful one. First of all polar explorers he illuminated the winter quarters by acetylene gas, banishing the bugbear of perpetual darkness.

He personally chose and tested the rations and their packing. For the long southern journey he introduced Manchurian ponies, also for the first time, to draw sledges so designed that when the animals failed they could be hauled by men. He made use of improved methods of camping, of handling the sledges, and of marking the track, with the result that he accelerated the rate of sledge travelling to a surprising extent.

Starting from the same base as the *Discovery* expedition, he laid out the route far enough to the east to escape at the outset the worst crevassed area of the Barrier. These and a myriad improvements in detail enabled Shackleton and his three companions to reach the former "Farthest South" in twenty-three days instead' of thirty-three. Six days beyond this point he discovered the Beardmore Glacier stretching gently upwards like a broad highway to the Pole through a vast valley cleft in the mountain wall, Fair to see, the road was rough to travel; the last pony was lost in the waste of jagged ice-masses, and by their own strength the men dragged their sledges 120 miles from the level Barrier up to the edge of the plateau nearly 10,000 feet above the sea, and for 180 miles farther until within 100 nautical miles of the South Pole. At intervals of about 100 miles on the outward way the loads were lightened by leaving a supply of food and fuel just sufficient, if no delays occurred, to carry the party back to the depot behind.

The only clue to guide the returning travellers to those icy molehills was the faint track left by their outward-bound sledge in the snow and a few snow-pillars. To miss any one of them would mean certain death. In order to compensate for food consumed during unforeseen delays rations were cut down to the very minimum that could support life; but nothing could be done to make up for loss of time as ultimate safety depended on returning to the base before winter forced the *Nimrod* to leave. Shackleton held on his southward way until the last possible moment; then he turned. He had solved the problem of reaching the Pole, and but for the delays which no foresight could avoid he would have reached it. The return journey was a grim and doubtful race with Death; but from point to point the race was won. Each tiny cache of food and fuel was found and reached just as the supplies carried gave out; once after the party had been forty hours without a meal. Never was safety plucked from danger by so close a call.

In spite of all difficulties the route was mapped so completely and accurately as to furnish a trustworthy guide to the intrepid party which slogged on to the Pole itself three years later. Shackleton planned and his comrades performed two other unprecedented feats on this expedition:

the ascent to the active crater of Mount Erebus, 13,000 feet above the frozen sea, and the attainment of the South Magnetic Pole.

The labour of the expedition over, Shackleton carried out two others almost equally Herculean, the writing and publication in eight months of two, great volumes which bore no sign of haste in literary style or artistic illustration, and a lecture tour of the world with the first kinematograph film ever taken in the Far South, to raise funds with which to pay off the guarantors whose support had made the expedition possible. Shackleton's ambition grew with his success. His fertile brain conceived new plans and devised new methods to achieve them, no full account of which has ever appeared. The Imperial Trans-Antarctic Expedition was launched in the unforeseen eclipse of peaceful enterprise just as the shadow of the Great War swept over the world.

Shackleton, starting only when his offer of the ship to the Admiralty for war purposes was declined, hoped to land in the south of the Weddell Sea, to sledge across Antarctica to the South Pole and on to the Beardmore Glacier, where a party sent out by him to the Ross Sea had deposited stores to ensure his safe arrival at McMurdo Sound.

No plans could have been better laid; but the tide of Shackleton's fortune had turned. The *Endurance* was nipped in the ice of the Weddell Sea and drifted slowly northward to destruction. As in a Greek tragedy disaster after disaster fell upon the expedition as the leader waited helpless but undaunted on the splitting and dwindling floe, and as he battled later making his boldness wait on caution in the small boats amongst the huge seas of the wildest ocean in the world. His spirits rose to meet the shocks of doom and in the end he brought back all his men, shaken indeed, but not shattered by the ordeal.

His first expedition established Shackleton as an explorer unequalled in the audacity of attack; his second revealed him as a man at grips with Fate, prudent and cautious in retreat. In both he “was ever a fighter." Strongly contrasted qualities met in Shacketon's many-sided personality, reflecting his descent from quiet, learned, solid Yorkshire Quakers on the one side, and from quick-witted, impulsive, and pugnacious Irish ancestors on the other. Physically he was strong, mentally he was alert, morally he was sympathetic with all that is good; but his success and his fortitude cannot be ascribed to devotion to any one ideal. Many aims attracted him, but perhaps his strongest motive was a passion for self-expression in adventurous action: in all his strivings he looked high and "hitched his Wagon to a star." His daring often looked rash, but it stopped just short of recklessness. His caution only blossomed in the field of danger, and then it concerned others, not him self. In moments of crisis his decisions were so swift and his action so prompt that he seemed to act on some impulse of intuition, though really it was quick observation and rapid reasoning that lit his way. His mind was so balanced as to be sensitive to the faintest indications, and at its highest tension his choice was always instantaneous and right.

Shackleton loved poetry and he had the power of a poet to imagine all the possible eventualities of any situation; he had also the patient precision of a storekeeper in providing for each emergency that he foresaw. In everything he did he overflowed with a self-confident cheeriness which carried every one with him wherever he would have them go. No leader was ever more loved and respected by his followers; their devotion owed nothing to authority, but everything to the warmth of genuine comradeship which made his will the will of all who were with him.

When dangers thickened round him the lighter side of his nature faded and the deep-seated greatness of his character rose and possessed him entirely. This is the aspect which he bears in the noble statue now to be unveiled, all fortitude and patience; but we, the friends of his dynamic youth, can animate the bronze with the memory of his keen and humorous eye, his vibrant voice, his winning smile and rollicking laughter, all speaking to us of his quixotic generosity, his delicate tact, and his overflowing good fellowship.

THE MARQUESS OF ZETLAND:

With admirable lucidity and with a few strokes of a master brush Dr. Mill has painted for us a vivid picture of the achievement on which, as he has told us, Shackleton's claim to fame must ultimately rest. And I ask myself what was the outstanding feature of the achievement the journey of 1909, when Shackleton with three companions forced his way over 366 miles of hitherto un trodden ice and snow from the southernmost point reached under the leadership of Scott in 1902, to a spot 100 miles only from the Southern Pole?

To me, looking at the matter with the eyes of a layman having no personal experience of Arctic travel, it was the extent to which the planning, the financing, the organizing, and, in a sense, the final execution of the project were the work of a single man. There was no committee of organization, no public appeal for funds, no Government help just an individual fired by a great ambition working away doggedly and with unflagging optimism to render possible the attainment of a distant but alluring goal. “I have had three years' struggle and have tackled seventy-odd rich men,” he wrote on 4 February 1907; and on February 11, with no fanfare of trumpets, but unassumingly at a dinner of the Kosmos Club, his plans for the Conquest of the Antarctic were made known. That seems to me to have been typical of the man-strong, confident, self-contained.

It is significant, surely, that on Shackleton's mere promise to make good the cost out of the proceeds of writings and lectures which must obviously be dependent on success, shrewd, hard-headed men of business should have been found willing to guarantee the sum required; significant too that fellow travellers, familiar with the trials and the risks of Polar exploration, should have been ready to follow without question wherever Shackleton chose to lead.

Here, clearly, was a man of parts inviting something more than passing notice.

He was, in fact, a striking example of a type: a type which has sent forth into all lands pioneers in exploration, in commerce, in industry, and in administration; an adventurous and a ruling type; a type which seems always to have flourished on British soil; one without which Britain would have remained but an insignificant island and Greater Britain would never have come to be.

Let us then, since not the least of the objects which we have held in view in raising this memorial to him has been to direct attention to the example which he provides for future explorers of our race, devote a few minutes to a study of his character and personality.

As a boy Shackleton was surely not the sort that you would have expected to break away from the family tradition which suggested, for the most part, a studious and quiet existence. He loved music and poetry and he was markedly religious, characteristics which at first sight might be expected to find a more congenial setting within a study's walls than amid the harsh and crude surroundings of the world's unbeaten tracks. Yet such characteristics are indications of emotional deeps where lie dormant forces which, when rendered active by stimulus from without, are capable of influencing profoundly human action, notable among such forces, idealism and ambition.

History abounds with examples of such cases. As a young man, Evelyn Baring, afterwards Earl of Cromer, now known throughout the world as the Maker of Modern Egypt, was conscious of not the smallest tittle of ambition, and it is quite likely that, but for his chance meeting in the island of Corfu with the lovely English girl who fourteen years afterwards became Mrs. Baring and subsequently Lady Cromer, he would have lived and died unknown. Ambition came to the surface of the emotional deeps in Shackleton's nature as the result of a similar stimulus from without; for in 1897 he met Miss Emily Dorman, who seven years later became Mrs., and afterwards Lady, Shackleton.

"In my bunk that night," wrote Mr. Dunsmore, after a talk with Shackleton on board the Flintshire, the ship on which the two of them were serving in 1897, "I felt convinced that the ambition of that man's life was to do something worthy, not only for himself but for her." Given ambition, it is curious how often the particular direction in which it carries a man seems to be the result of accident. It was a whole series of apparent accidents that took Evelyn Baring to the shores of Egypt; similarly Shackleton's choice of a seafaring career which led him in due course to the frozen and inhospitable wastes that rise from the depths of the Antarctic Ocean, seems to have been purely accidental.

All that we know is that at an early age an illustrated book of Arctic travel, Mr. C. F. Hall's 'Life With the Eskimo,' came into his hands and that thenceforth he read all that he could obtain about Polar exploration; and further that contrary to his father's hope that his son might follow in his footsteps as a doctor, young Shackleton at the age of sixteen went to sea.

I am going to indulge in a generalization; but before doing so may I remind you how deeply the characteristics of which I have been speaking were implanted in young Shackleton's nature. Life before the mast in a sailing clipper was little calculated to cultivate the finer feelings; yet it did little to choke the idealism which permeated young Shackleton's being. He was afraid that he would be laughed at when he said his prayers, but “the first night I took out my Bible,” he wrote, “they all stopped talking and laughing, and now every one of them reads theirs. . . " And the illimitable expanses of sea and sky stimulated the poetic imagery which welled up from his innermost being - "Many a painter would have given half of what he possessed to have been able to catch the fading tints of the red and golden sunset we had last night," he wrote when still in his 'teens, and "if you wish to see Nature robed in her mantle of might, look at a storm at sea; if you want to see her robed in her mantle of glory, look at a sunset at sea."

Now for my generalization, which is this, that such natures, if they are to function to advantage, must pass through the crucible of stern and great events. It is in crises that such men attain to their full stature; in the bearing of responsibility and the welding of authority that they rise to their true heights. In the routine life of a humdrum everyday existence their talents are apt to run to waste. Shackleton provides us with a striking example of the force of this generalization. In England, when not engaged in planning and organizing his Polar Expeditions, his restless spirit sought distraction in divers enterprises as Secretary of the Royal Scottish Geographical Society, as a candidate for Parliament; in journalism and in business, often of a highly speculative character, in which he was always going to, but never did, make a fortune.

Let it be said in passing that his ambition to make a fortune was the outcome not of any sordid love of money, but of his vivid imagination which caused him to revel in the prospect of endowing his wife and family with affluence and of wielding the power which the command of riches is generally supposed to give. In actual practice he was reckless in his generosity. Appreciable sums of money which flowed in as a result of his lectures in New Zealand on the eve of his expedition in 1908, sums badly needed for the purposes of the expedition itself, he devoted to local charities.

This was an episode of repeated throughout his career. He had, in fact, much of the spirit of the gambler; his delight lay not in the actual acquisition of wealth but in the prospect of climatic throws with fortune and in the quixotic spending of such riches as happened to come his way.

The prospect of fame, and possibly of fortune, was not without its influence in directing Shackleton's gaze in the first instance towards the Southern Pole; but unless I have gravely misread his character a subtler and profounder Influence was at work, namely, an idealism of a high order: the idealism on whose wings the spirit of man seeks to rise superior to the limitations of his material environment, finding its reward less in the fruits of victory than in the actual process of attaining; an idealism akin to that which led Mallory and Irvine to a glorious death far from the haunts and the huckstering of men, high on the topmost pinnacle of earth.

Thus it was, as we should expect, when warring with the stark realities and the constant dangers of the Antarctic that Shackleton was at his best. It was when thus tested that his powers of leadership, his resource and judgment, his courage and endurance came into play.

It is a remarkable tribute to his qualities that he never lost a life in any one of the expeditions that he led. Let those who would assess this fact at its true value study the amazing retreat of the members of the Trans-Antarctic Expedition after the loss of the *Endurance* in the ice in 1915. It may be argued that for a single moment and that a fatal one-Shackleton's judgment failed him; but no one can harbour feelings other than those of the highest admiration for the heroic qualities which he displayed during those long and nerve-shattering weeks when he and his party drifted helplessly on the ice floe; during the desperate battle with the waters when, their very world cracking and dissolving beneath them under the influence of the warmer waters into which they had drifted, they made in three small boats across 100 miles of turbulent ocean for Elephant Island; during the seventeen interminable days of strain and terror when with five companions he fought his way in one small boat across 800 miles of water from Elephant Island to South Georgia; and finally during that crowning incident of courage and endurance when With two companions but without guides or maps or resting-places, he set forth to cross the snow-clad and untrodden range of mountains which lay between him and help in the shape of the small population of a whaling station. One would have to search long, indeed, to find in the annals of any Northern Race a Saga of the Sea more thrilling in its incidents, more moving in its emotional appeal, than that set forth with so fine an understanding in the pages of Dr. Mill's biography.

Such is the man whose memory we seek to perpetuate as an example to his race, by means of the statue of which I now invite you, Sir, as President of the Royal Geographical Society, to accept the custody.

ADMIRAL SIR WILLIAM GOODENOUGH (President R.G.S.):

Lord Zetland: Before all else may I, at the request of Lady Shackleton and her family, oïïer to you and your Committee their sincere thanks for all your care in raising and administering the fund which has led to this ceremony. It is with feelings of pride and reverence that on behalf of the Royal Geographical Society I accept from you and the Committee of the Shackleton National Memorial the charge of this statue. It will serve to remind us and all who pass by, of a man who had he lived in the golden age would have been a Drake or a Raleigh; of one among those men whose names encircle this hall as they themselves encircled the world; of one to whom adventure was an inspiration, danger an incentive. Imbued with the true spirit of Polar Exploration he counted no cost to himself, but was ever mindful of his companions; a great leader of men, beloved and trusted by his followers.

SIR ERNEST H. SHACKLETON.

How I began by Sir Ernest H. Shackleton
An interview by A.B. Cooper

SIR, ERNEST H. SHACKLETON is young enough to remember how he began without any very great effort of recollection. It was only yesterday that he did begin, and for all his fame he is still little more than a big schoolboy. On the other hand, however, it is doubtful whether he ever was just like other boys, I had not the privilege of knowing him in the days of his youth, but it is possible to reconstruct a man's boyhood out of the material of his manhood I should say that Sir Ernest Shackleton was rather an odd boy who, in spite of his adventurous nature and the spirit of romance that was in him, loved a book better than a bat, solitude better than a crowd, his own companionship better than a mob of other lads; a boy who kept his own counsel, never except for brief moments let his thoughts show in his face, or make themselves audible on his tongue a grim, determined, headstrong fellow, of whom the discerning would prophesy great things and the undiscerning miserable failure.

Everybody knows by this time that Shackleton was a Dulwich boy and Dulwich is mighty proud of him.

Perhaps it is a good thing that schoolmasters have not the prophetic gift, that it is not given to them to peer into the future and see what the boys they are teaching, and squashing and encouraging and swishing by turns, are destined to become and to do in the world I say perhaps it is a good thing, because otherwise some boys would be spoiled by too much care and others ruined by too little. The elemental forces of school life are a great unconscious training for a boy and from these blind and semi blind, forces there emerge independence, grit, initiative, and once in a while, greatness.

Sir Ernest naturally dissents from the notion which has got abroad and has appeared in countless public prints that he was a bit of a dunce 'at school' and that, whatever credit and merit Dulwich may take to itself for his later achievements, it had no reason to be proud of him as a scholar. This idea got abroad from an interview with a master which was not correctly reported.

“I wonder how many' boys would leave an impression upon the record of their school, or upon the memory of their teachers, if they entered the school at twelve and left the moment they had turned sixteen ? ”

Said Sir Ernest to me. “I was in the Modern Lower Fourth 'when' I left.”

“I went to sea when I was sixteen. Yes, it is quite true that my father was against it very much against it but it is not true, as has also been asserted that I ran away from school and from home; I simply did not return to school when the next term started, but went to sea instead. I will say this, that my father thought to cure me of my predilection for the sea by letting me go in the most primitive manner possible, as a 'boy' on board a Sailing-ship at a Shilling a month! Twelve shillings a year for working hard, facing all kinds of weather, 'beating round the world against every wind that blew that's the sort of school I swapped Dulwich for."

“But you' found it a good school” I asked.

“Yes, I did,” said Sir Ernest.

“It taught me things' and that is what Dulwich, with all its good points, had not done. I never learned much geography at school, for instance my first voyage taught me more geography than I should have learned had I remained at school to the age of eighty. School geography consisted in my time there may be a change for the better today in names of towns, lists of capes and bays and islands. Literature, too, consisted in the dissection, the parsing, the analysing, of certain passages from our great poets and prose-writers. But, though it is hardly believable, I learned more of literature in a year at sea than I did in half a dozen years at school. I seemed to get at the heart of it then, to see its meaning, to understand its message, and find some degree to catch its spirit."

"You were a reader, then?" I asked, thinking that men of action were not always men of books.

"Always, my reading took two main lines travel and exploration, to which you may add history, though you may consider that a line of itself, and on the other hand, poetry.

Yes, I read poetry as a boy and I've read poetry ever since. I greatly love and admire Browning. He is a fine spirit. I like his optimism, his note of 'Never say die, the grand way in which he faces the future, his outlook upon the world. Yes, I think Browning is great. I do not like all poetry. I may read it, but it does not appeal to me. Nor do I like all the poems even of a favourite poet. Certainly Browning pleases me almost all the time, but I would pick and choose from Tennyson.

I am sure of this, too, that poetry is good for boys to read, and I think boys should be encouraged to read poetry, and teachers should be very careful for poetry for all time an imposition not to spoil their taste by making it a task and I read Motley's History of the Dutch Republic when I was at sea, and its fascinating story of the way that little nation became a great naval power and a great colonising race fired my imagination. Prescott I read and, indeed, most of the historians I do not know how I came to be drawn to the reading of history I mean, a certain type of history. I was never much taken with records of battles and sieges, and dynasties and dethronements, but of the history of a people struggling to be free, of adventurous. Nations who sent their mariners into unknown seas, and the history of colonisation and exploration, I was always very fond. Then of course, the unexplored parts of the world held a strong fascination for me from my earliest recollection. I was more or less acquainted from my earliest youth with all the problems of exploration, whether it was in Central Africa, Central Australia, Tibet, the North and South Poles, or New Guinea and Borneo.

'I must say' that the North Polar Region was the part of the globe towards which my mind was most strongly drawn read everything I could lay my hands upon which had for its subject Polar exploration. One of my favourite books then, as it is to-day, was Dr. Hall's Life with the Esquimaux. I am surprised that no one thought to republish that fascinating polar classic. Then, of course, Captain Kane, McClintock, Franklin, Markham, and all the old explorers of the Arctic, were first class heroes to me, and I not only read their books, but remembered what I read. I may truly say that long before I went to the Antarctic myself I knew everything and everybody that had been before me, to either end' of the earth's axis, and not because I had set myself to learn it had that been the case, probably I never should have done it but simply because my mind naturally sought and assimilated details of this kind.

“And where,” I asked, “did you first get the notion that you yourself would join this gallant band and become an explorer yourself?”

“I think it came to me,” replied Sir Ernest, “during my first voyage. I do not say that it definitely formed itself in my mind as a purpose that came later but on that occasion I felt strongly drawn towards the mysterious South. During that voyage, which constituted one of the stiffest apprenticeships, surely, that ever a boy went through, we rounded Cape Horn in the depth of winter. It was one continuous blizzard all the way, one wild whirl of stinging sleet and snow, and we were in constant 'peril of colliding with icebergs or even of foundering in the huge seas. Yet many a time, even in the midst of all this discomfort, my thoughts would go out to the southward, across that great expanse of southern sea, the loneliest tract of ocean in the wide world, the region which seemed to have been especially guarded against the approach of man by the great ice-barrier.

"I DREAMT I WAS STANDING ON THE BRIDGE IN MID-ATLANTIC AND LOOKING NORTHWARD."

But, strangely enough, the circumstance which actually determined me to become an explorer was a dream I had when I was twenty-two. We were beating out to New York from Gibraltar, and I dreamt that I was standing on the bridge in mid-Atlantic and looking north-ward. It was a simple dream. I seemed to vow to myself that some day I would go to the region of ice and snow and go on and on till I came to one of the poles of the earth, the end of the axis upon which this great round ball turns. I am sure my thoughts then turned to the North Pole, because it was more in men's minds and had been for generations than the South Pole. But, you see, my thoughts were wrong, though the dream in its essence was right. After that I never had any doubt that sooner or later I should go upon a polar expedition."

"After my first voyage," continued Sir Ernest, "my parents wanted me to stay at home and give up the sea. Had I found it all that my boyish imagination had conjured up? I must confess I had not I do not think a boy ever does. I think if I had hated the sea I should have still stuck to it. But I didn't hate it, although I found it a hard life. I went out again round the Cape of Good Hope to India, Mauritius and Australia. This was the roughest voyage of my experience.

By The Lords of the Committee of Privy Council for Trade.

Certificate of Competency

as

SECOND MATE.

To Ernest Henry Shackleton

Whereas it has been reported to us that you have been found duly qualified to fulfil the duties of Second Mate in the Merchant Service, we do hereby, in pursuance of the Merchant Shipping Act, 1854, grant you this Certificate of Competency

By Order of the Board of Trade.

this 30th day of August 189[illegible]

Countersigned, Assistant Registrar General

Ingram B Walker { One of the Assist Secretaries to the Board of Trade.

Registered at the Office of the Registrar General of Shipping and Seamen.

CERTIFIED TO HAVE PASSED IN COMPASS DEVIATION 189 REGISTRAR GENERAL OF SHIPPING & SEAMEN LONDON

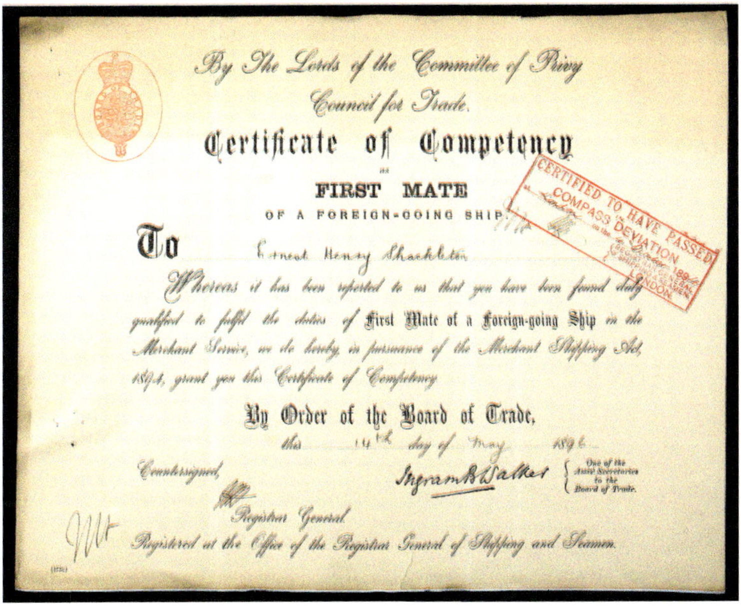

By The Lords of the Committee of Privy Council for Trade.

Certificate of Competency

as

FIRST MATE

OF A FOREIGN-GOING SHIP.

To Ernest Henry Shackleton

Whereas it has been reported to us that you have been found duly qualified to fulfil the duties of First Mate of a Foreign-going Ship in the Merchant Service, we do hereby, in pursuance of the Merchant Shipping Act, 1894, grant you this Certificate of Competency

By Order of the Board of Trade,

this 14th day of May 1896

Countersigned, Registrar General

Ingram B Walker { One of the Assist Secretaries to the Board of Trade.

Registered at the Office of the Registrar General of Shipping and Seamen.

CERTIFIED TO HAVE PASSED IN COMPASS DEVIATION 1896 REGISTRAR GENERAL OF SHIPPING & SEAMEN LONDON

Sometimes a vessel seems to dodge all the storms and sometimes to encounter them all. The latter was our experience. We had a rough time, a time which you can imagine for yourself when I tell you that the vessel I sailed in, a sailing-ship was dismasted twice. On that occasion I was away from England two' years."

"Meanwhile, I had been studying and I passed for second mate and shipped on a tramp steamer to China and Japan. Whilst I was at sea I continued my studies in navigation and seamanship. No, I never asked a great 'deal of help from anybody. I have always believed in doing things on my own, as the saying is, so I just worked away and picked up my knowledge wherever I could find it.

Personally, I find that all experience helps and that' there need be very little waste in life."

"On my return I passed for mate and shipped as second officer on a big tramp of eight thousand tons which was sailing for the East, and whilst at Singapore I passed for captain. Since then I have had a mixed sort of existence. I took a turn at journalism in London; I went on a troopship to South Africa; I wrote a book; then I heard that Captain Scott was wanting volunteers for a South Polar Expedition, went to see him, and was after a long wait accepted, and there you are. My dream had come true, except that I went to the South instead of to the North."

"I TOOK A TURN AT JOURNALISM."

There are some things which this almost fiercely modest man is a little shy of revealing, although they are greatly to his credit and lie very near his heart. One of these is that he is not only a lover of the poetry of the Poets, but that he is 'a' poet himself. When I visited him, as I did on several occasions, in his cabin like office at the lower end of Regent Street, he put into my hands a beautifully bound and reproduced facsimile of the "South Polar Times," and as a special favour granted to the readers of THE CAPTAIN he gave me permission to copy from it a poem written in the great darkness of the Southern night, probably the only poem in the World that was ever penned amid Antarctic snow and ice. It is entitled

'To the Great Barrier'

Mother of mighty icebergs these Kings of the Southern Seas,

Mystery yet unfathomed, though we've in full our fees.

Eyes strained by ceaseless watching, when the low grey fog doth screen.

Your walls from our aching vision and the great grim giants wean.

Away from your broad white bosom, where for aeons untold is laid.

Each yearly tribute of fallen snows, that this wonderful plain has made.

We have felt, more than seen, the danger, close ahead of our long jib-boom.

But a turn of the icy wheel has made for us a more sea-room.

We have sailed from your Farthest West, that is bounded by fire and snow.

We have pierced to your Farthest East, till stopped by the hard-set floe.

We have steamed by your wave-worn caverns, dim, blue, mysterious halls,

We have risen above your surface, we have sounded along your walls.

And above that rolling surface' we have strained our eyes to see,

But league upon league of whiteness was all that there seemed to be.

Ah, what is the secret you're keeping, to the Southward, beyond our ken?

This year shall your icy fastness resound with the voices of men?

Shall we learn that you come from the mountains? Shall we call you a frozen sea?

Shall we sail to the Northward and leave you, still a Secret for ever to be?

Ireland, as well as England, is proud of the Antarctic explorer, because he is a scion of an Anglo-Irish Quaker family which first settled in Ireland in 1727.

The first settler, who came from Yorkshire, was Abraham Shackleton, who established a famous school at 'Ballytore', Co. Kildare, at which Edmund Burke, the great 'orator and parliamentarian, received his early education. One of Sir Ernest Shackleton's old teachers at Dulwich lately remarked that Shackleton was generally looked upon after he left school as a rolling-stone who would probably gather no moss.

"But," added the master, with a twinkle in his eye, "the event proves that though he may not have' gathered much moss, he has at any rate rolled further than anyone else."

And so he has. Experience, joined to personal dash, took him three hundred and Sixty nautical' miles nearer the South Pole than Scott; it enabled some of his party to reach the Southern Magnetic Pole of the earth, and to get within one hundred and eleven miles of the South Pole itself; to discover eight new mountain chains, and survey one hundred mountain peaks to ascend Mount Erebus, the most Southern Volcano in the world, which rises 13,120 feet, and to descry a new coast and still another far off mountain range, and even, in this vast Southern Continent, to find coal measures which prove that at one time, so far from these polar regions being clothed in a great ice-cap, they were covered with a tropical jungle.

A. B. Cooper

From the John Rickinson's Louis Rickinson Collection

Louis Raphael Rickinson (born 21 April 1883 (Lewisham), died 16 April 1945 (Newbury, Berkshire)) was an English marine engineer and served as Chief Engineer on shackleton's Endurance Expedition 1914 -1916 , for which he was awarded the Silver Polar Medal.

Louis Rickinson with his wife and son.

Louis kept an autograph album and opposite you will see a full page of signatures of those on the Quest Expedition.

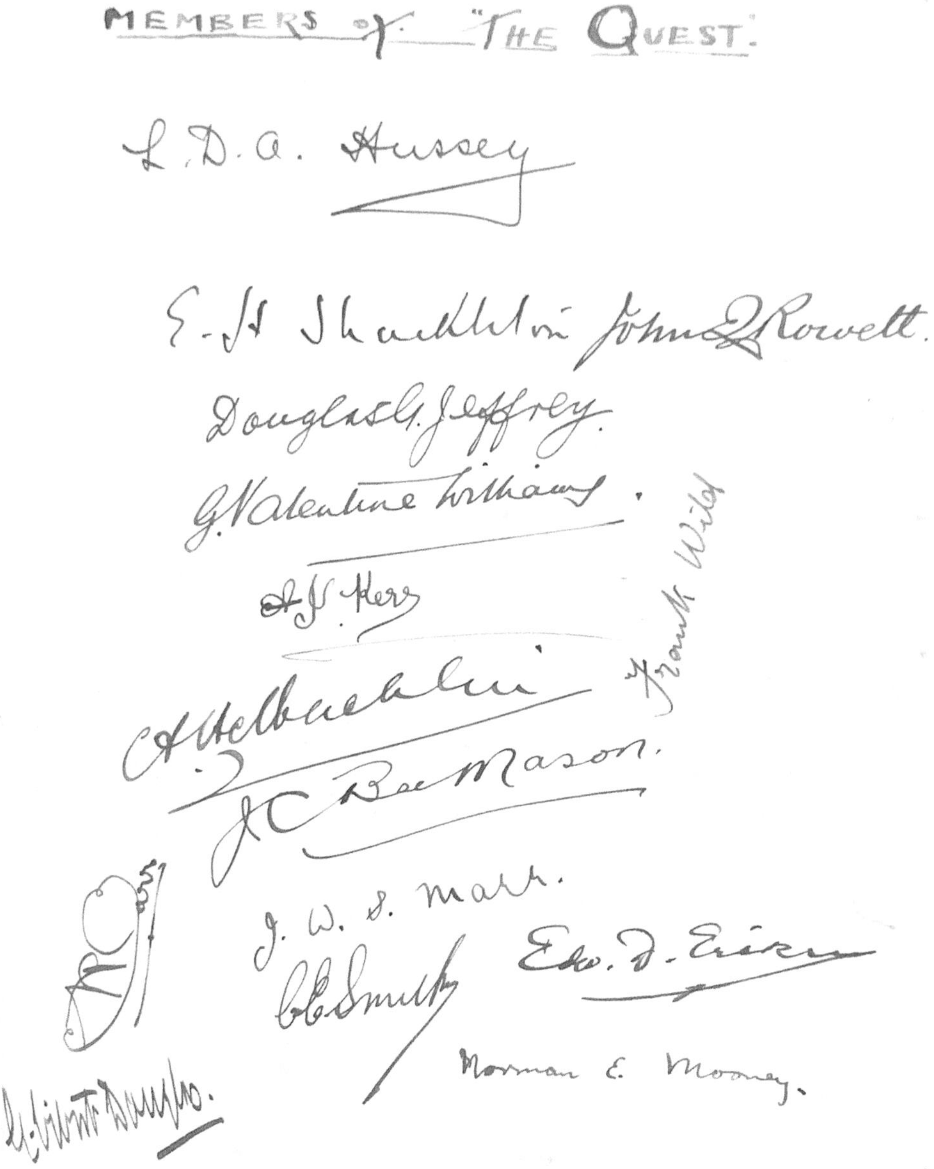

The Quest Expedition page from the Louis Rickinson autograph album

Home

Drawn from memory by Louis Rickinson of his time living in the upturned boat, during the Endurance Expedition.

Linocuts on Shackleton
by Bill Payne

* 9 7 8 1 9 0 1 0 3 7 2 1 0 *

BV - #0124 - 011225 - C23 - 210/140/14 [16] - CB - 9781901037210 - Gloss Lamination